AFTER CONTACT

The Human Response to Extraterrestrial Life

AFTER CONTACT

The Human Response to Extraterrestrial Life

ALBERT A. HARRISON

PLENUM TRADE • NEW YORK AND LONDON

Library of Congress Cataloging-in-Publication Data

On file

ISBN 0-306-45621-4

A Division of Plenum Publishing Corporation
233 Spring Street, New York, N.Y. 10013-1578
http://www.plenum.com

10 9 8 7 6 5 4 3 2 1

Printed in the United States of America

To the women in my life:
Mary Ann, Kathy, and Alycia Marie

PREFACE

Recently, while rummaging through family memorabilia, I uncovered my earliest documented work on extraterrestrial intelligence. In a letter to my mother dated three days short of my tenth birthday, I described in great detail entering a drawing contest sponsored by the host of a local television show recorded for posterity as "Doodle with Galligan." The contest involved sketching people from other planets. The letter was richly illustrated with my rendition of Galligan's example and my own contest entry.

Galligan's extraterrestrial had a body reminiscent of an elongated egg, three otherwise conventional legs and feet arranged as a tripod, no neck, and a smiling human face. It was encased in a tight space suit that had external earphones attached to an antenna of sorts, not unlike the headphone radios used by some joggers today. My extraterrestrial was even more human in appearance. It differed from the average person only in having nine fingers on each hand (inspired, no doubt, by the genetic anomaly of one of my fellow fourth graders) and bulky space suit (inspired, perhaps, by those of the first lunar astronauts as drawn for the children's newspaper *The Weekly Reader*). Both Galligan and I must have envisioned these people from other worlds landing on Earth; otherwise, there would be no need for space suits. Since it lacked arms, Galligan's alien showed much more originality than mine. Despite my best efforts, someone else must have won the cornucopia of pens, pencils, brushes, crayons, inks, and paper that was valued at the equivalence of 30 weeks' allowance. Both Galligan and I went on to other things, in my case only temporarily.

Many years would pass before I again gave sustained thought to life in outer space. But as my school and then college years slipped by, scientists made a succession of discoveries hinting that many intelligent life-forms were scattered throughout our Milky Way galaxy and raising the possibility of establishing contact with at least some of them by means of microwave radio.

In the late 1970s I began collaborative writing on the psychological and social dimensions of life in space—human life in space. There were many opportunities to discuss this work, including a meeting of the Society for Applied Anthropology held in Reno, Nevada in 1986. Whereas I talked about the role of anthropologists in studies of human spaceflight, other speakers discussed extraterrestrial intelligence, and I was hooked. Over the following years I read voraciously on the topic and met with a group of anthropologists, philosophers, scientists, engineers, science fiction writers, and students who hold an annual three-day symposium on the topic. Led by anthropologist Jim Funaro, this group—which goes by the straightforward name CONTACT—taught me much and became a testing ground for many of my ideas. In the late 1980s I decided that some of these ideas should be put in writing; this led to two articles in *Behavioral Science;* one, with Alan Elms, discussed the role of psychology in the search for extraterrestrial intelligence (SETI), and the other presented a conceptual framework for hypotheses about extraterrestrial organisms, societies, and organizations of societies, or "Galactic Clubs." These articles were the seeds of the present volume.

After Contact: The Human Response to Extraterrestrial Life is organized into three parts. Chapters 1 through 3 describe the rationale for a belief in extraterrestrial civilizations and for proposed search procedures. In these chapters we will see why, early on, SETI scientists encouraged the involvement of psychologists, anthropologists, sociologists, and other behavioral scientists, and we will explore the role of these fields in the search process. Because psychiatrists and psychologists are showing an increased interest in UFO sightings and alleged human abductions by aliens, we will consider the UFO controversy and examine a few principles of human psychology that together may explain sightings and abductions. We will also see why UFO experiences do not constitute acceptable evidence of extraterrestrial intelligence, and will suggest that the great gulf between SETI scientists and UFOlogists (people who study UFOs) is appropriate.

The second section of this book, chapters 4 through 7, speculates about extraterrestrial life-forms and societies. Is our ignorance of life "out there" so profound that we cannot begin to imagine what "they" will be like? Or are there some justifiable assumptions and useful tools that will allow us to make intelligent guesses as to their nature? Like the astronomers, physicists, and exobiologists who have speculated on extraterrestrial life-forms, I adopt the second premise. The possibilities are not endless; there are constraints at least on the range of life-forms with whom we might establish radio contact. During earlier centuries scientists sought to establish that the laws of physics apply throughout the universe, and for the last hundred years or so have sought to discover if there are universal laws of biology as well. This section represents an extension of this effort into behavioral science.

We will explore the view that the scientific assumptions that serve us well on Earth will also apply to our speculations about alien life-forms. As we develop hypotheses about extraterrestrial biological and social entities, we will remain within the framework of science and view the evolution of life and civilizations as orderly processes that proceed within broad natural limits.

James Grier Miller's Living Systems Theory (LST) is our point of departure. LST has broad applicability over time and across species and cultures. It stresses the continuities among the physical, biological, and social sciences, and it provides a simple framework for disassembling and analyzing, in identical terms, biosocial systems of different size and complexity. We must rely on a simplified version of LST involving three systems levels (organism, society, and supranational systems) and two basic processes (matter–energy and information processing) to organize current thinking about extraterrestrial intelligence. Many of the hypotheses that astronomers, physicists, and biologists have formulated about intelligent life in the universe are consistent with LST, and the theory should be useful for future comparative studies of terrestrial and extraterrestrial life-forms and civilizations. Chapter

4 describes LST, and chapters 5 through 7 apply the theory to organisms, societies, and supranational (or interstate) systems.

Chapters 8 to 11 move us to slightly safer ground as we explore hypotheses about the human reaction to contact. (The ground is firmer because we already know something about one of the two partners in the transaction, humans.) We begin with a discussion of the social and psychological dynamics that are likely to affect our first impressions. Most likely, our impressions will be based more on media accounts and on individual and group psychology than on hard data about the aliens themselves. Whatever their basis, first impressions will be crucial, because they will set the stage for our initial response and will set into motion forces that could color or even shape relations between the two civilizations. We then turn to the possible initial or short-term impact of alien–human contact.

Here, we will draw on historical precedents where large numbers of people believed that we had contacted—or had been contacted by—extraterrestrials. These prototypes include the alleged sighting of life on the moon, the belief that Mars was the home of a dying population, misinterpretations of quasars and pulsars as extraterrestrial beacons, and, of course, Orson Welles's famous *War of the Worlds* broadcast. Under the more likely contact scenarios, the *War of the Worlds* panic may be the least useful model. Moreover, many lines of defense keep people from panicking under stress, and evidence of this is found in a careful study of human reactions to the *War of the Worlds* broadcast. The next two chapters discuss, in turn, how we might proceed if we conclude that active communication is necessary or desirable, and what the possible long-term effects of contact on our people and societies might be. I take a close look at the assumptions underlying the belief that contact with an advanced civilization will bring us immense benefits, and reflect on the adage "Be careful what you wish for, because you might get it."

The final chapter reviews the journey that humankind has taken from determining the number of stars in the galaxy to considering seriously how Earth's problems might be solved with the aid of extraterrestrials. This chapter highlights the themes of the book, summarizes why we have need to be patient but have cause to be optimistic, and explains why the debate over extraterrestrial life may be extremely difficult to resolve.

ACKNOWLEDGMENTS

Many people have helped inspire and shape this project. I am particularly indebted to my friends and colleagues who regularly attend meetings of the group of anthropologists, scientists, and science fiction writers known as CONTACT. Jim Funaro, Poul and Karen Anderson, Greg Barr, Paul Bohannon, Ben Finney, Martyn Fogg, Joel Hagen, Dick Haines, Ken Koenig, Don Marshall, Chris McKay, Gerald Nordley, Reid Riner, Don Scott, Seth Shostak, Bob Tyzzer, and many others have been sources of inspiration, ideas, and helpful criticism over the past ten years, and they will find some of their thinking reflected in this book. Mary Connors of NASA–Ames and my department colleague Alan Elms deserve special thanks for helping me get started, and I will forever be grateful to James Grier Miller, who not only created Living Systems Theory, the heart of this book, but who also encouraged me to publish speculative essays in *Behavioral Science*.

Several of my colleagues at the University of California at Davis, including Clay Ballard, Larry Coleman, Scott Gartner, Lyn Lofland, Don Owings, Bob Rerecech, and Clarence Walker, provided helpful comments on portions of this book—in some cases, substantial portions. I am particularly indebted to Frank Drake of the SETI Institute, Joel T. Johnson of UC Davis, and Allen Tough of the University of Toronto for reviews of the entire manuscript.

I want to thank my colleague Emmy Werner for bringing my work to the attention of Michael Hennelly, who provided unwavering encouragement and useful advice and brought my project to Plenum. This manuscript benefited in many ways from Karen Seriguchi's skillful editing. Most of all I am indebted to my partner, Mary Ann Harrison, who not only supported this incredibly time-consuming project but who accompanied me to meetings and symposia on "contact" and provided helpful comments on the technical and literary aspects of the entire work. All of my friends and colleagues who reviewed the manuscript have been good sports and gave generously of their time and expertise. Any errors or deficiencies that remain are my sole responsibility.

CONTENTS

7. SUPRANATIONAL SYSTEMS 173

8. FIRST IMPRESSIONS 197

9. INITIAL IMPACT 225

ONE

THE ENORMOUS CHALLENGE

As a final touch, just before initiating its interstellar journey, the technician thoughtfully added a friendly terrestrial "Hi" to the preprogrammed message.[1] For almost 25 years now, the greeting has hurtled through space at over 186,000 miles per second, the speed of light. Like the messages that people receive on their computers, it is digital, in this case a series of impulses intended to be arranged into 73 lines, each with 23 characters. Properly organized, the pixels, or "dots," will form a picture that includes representations of hydrogen, carbon, oxygen, and phosphorus atoms; chemical formulas for sugars and nucleic acids; DNA; and line drawings of a single human figure and a radio telescope antenna. Because the picture is of low resolution—1679 pixels, in comparison with the 380,000 pixels on the standard U.S. television screen—it is somewhat crude, and like a cloud or an inkblot, open to different interpretations.

Transmitted from Arecibo, Puerto Rico, on November 16, 1974, this proclamation of human presence helped mark the rededication of one of the world's largest radio telescopes. Loudspeakers enabled 200 scientists and guests to listen to the transmission, and the warbling, 169-second digital stream had an unearthly musical quality that brought tears to the eyes of many. The massive antenna pointed in the direction of the Hercules star cluster Messier 13 (M13), many light-years away.

As a practical matter, both the age of powered flight and the age of radio communication had begun in earnest slightly after the turn of the 20th century. Today, if they were seen from a great distance, the signs that ours is a spacefaring civilization would be sparse, but the signs that we are a radio-using or electromagnetically active civilization would be pervasive. Despite our few sprints to the moon, most human spacefarers remain in low Earth orbit, and there is little promise of an early breakout. It is highly unlikely that an extraterrestrial life-form (or, for simplicity, "ET") would first learn of our existence by encountering a human astronaut or finding equipment or debris in Earth orbit or on the moon. ET could discover one of our unmanned space probes that have completed their surveillance of other planets in our solar system and are now embarked on an endless journey to the stars. But compare the difficulty of finding these small objects with the ease of detecting a sphere of strong electromagnetic radiation, approximately 100 light-years in diameter, that is increasing at the rate of 186,000 miles per second and has already reached over 10,000 stars.[2] For the tiniest fraction of the energy cost of putting a small, unmanned satellite into orbit, the message that astronomers Frank Drake and his team launched from Arecibo passed the moon's orbit in less than 2 seconds, Jupiter's orbit in less than 35 minutes, and the outer limits of our solar system in 5 hours and 35 minutes.[3]

Since about 1960, Drake, Carl Sagan, and other scientists had been convinced that life is a frequent rather than rare occurrence in the universe, that at other places in our galaxy life has attained a form of intelligence that would be recognizable by us, and that the most plausible and efficient way for civilizations in different solar systems to find and communicate with one another is by means of radio.[4] The possibility of contact is basically a numbers game: given a target of 300,000 stars, perhaps a small handful will host civilizations that have attained our proficiency at radio astronomy, and with luck, one or two of these civilizations just might intercept our transmission.

Once intercepted, the physical properties of the message would help ET determine that the intercept was of intelligent origin, but decoding the message might take some time. The four dots followed by two dots that constitute "Hi" in International Radiotelegraph Code could pose difficulties, but the use of prime numbers to define the number of columns and rows would help the recipient arrange the pixels into a picture. We will never know whether a civilization in the vicinity of M13 will discover our message. Proceeding many thousands of times faster than our fastest spacecraft, the transmission will take 26,000 years to get there, so even a prompt reply would not reach us for 52,000 years. The true significance of the Arecibo transmission is its statement about our views of the universe, our place within it, and our willingness to find intelligent beings that are not like us.

Are there other electromagnetically active civilizations whose transmissions could be detected on Earth? By 1974, passive searches (that is, using radio telescopes to "listen" for signals from other worlds) had been under way for almost 15 years. That same year the National Aeronautics and Space Administration (NASA) conducted workshops on interstellar communication. The deliberations of these blue-ribbon panels at the NASA–Ames Research Center led to the conclusion that it was both timely and feasible to begin a significant SETI (Search for Extraterrestrial Intelligence) program. The technology was available, and large searches with great potential secondary benefits could be undertaken with modest resources.[5] In 1982 and again in 1991 the National Research Council reaffirmed the rationale and value of the SETI program.[6] Several searches, using technology that was far beyond our grasp in 1974, are under way today. These include major searches in Argentina, Australia, India, Italy, Russia, and the United States.[7]

SETI AS AN INTERDISCIPLINARY EFFORT

Although most clearly rooted in astronomy, physics, chemistry, and biology, SETI is a broad multifaceted effort with profound implications for all disciplines. Our focus is on the psychological, social, and cultural aspects of SETI: how fields such as psychology, sociology, anthropology, and political science can inform the search and help prepare humanity for different search outcomes. These fields can help us answer three fundamental questions that arise in discussions of SETI. How do human psychology and culture affect the form of the search? What, if anything, do our studies of terrestrial organisms, societies, and cultures suggest about organisms, societies, and cultures elsewhere in the universe? How will people respond to different search outcomes, including "contact," that is, definitive evidence that extraterrestrial life exists?

The Human Side of SETI

SETI is a human endeavor, based on human motivation and understanding. How do psychological, social, and cultural factors shape people's views about the likelihood of intelligent extraterrestrials and about the value of undertaking a search? How do human motivation and decision-making, organizational practices, and political realities shape the conduct of the search? Why is it that so many people—approximately 50 percent of the adult population of the United States—believe that extraterrestrials have already made contact with us, and what are the differences between UFOlogy, the study of unidentified flying objects, and SETI, the scientific search for extraterrestrial intelligence? We will consider such questions in this chapter and in the next two.

The Possible Nature of Extraterrestrials

Second, what might ET be like? What do we know from our knowledge of life on Earth that will help educate our guesses as to what our search might uncover? The task is formidable because thus far we have been able to study life-forms only on one planet, our planet, Earth. Yet, the NASA–Ames workshops recommended that we take serious stock of what we already know about intelligence and develop theoretical models that would allow extrapolation to extraterrestrial cultures.[8] Chapters 4 through 7 describe how James Grier Miller's Living Systems Theory,[9] a general theory that has proved useful for understanding a range of biological and social entities on Earth, may be suitable for organizing hypotheses and guiding speculation about life elsewhere in the cosmos.[10]

Human Reactions to Search Outcomes

How will people respond if SETI leads nowhere, or yields an equivocal outcome? How will people respond to definitive evidence that other solar systems host intelligent life-forms that have advanced to our level of technology or beyond? There is a negligible chance that if contact occurs, it will involve an actual meeting of humans and extraterrestrials, and a somewhat larger chance that we will find an alien space probe or other artifact. The most likely scenario for contact is that it will take the form of an intercepted radio transmission originating from a highly advanced society many light-years away.

If this occurs, what will be our first impressions? Will we panic? How should we proceed if we conclude that communication between our two civilizations is necessary or desirable? What are the likely short-term and long-term effects on our people and our societies? What role might humans play in galactic affairs? The questions are many, and the implications, notes Michael Michaud, "go far beyond normal scientific curiosity to the fate of intelligent life on Earth."[11] We focus on these questions in the final chapters of this book.

BEHAVIORAL SCIENCE CONTRIBUTIONS

Some of the disciplines that can help address such questions—psychology, sociology, anthropology, and political science—are in part biological sciences and in part social sciences. For sheer convenience, we will refer to these in the aggregate as the *behavioral sciences* but acknowledge that in some ways each of these fields is unique.

Physical, biological, and behavioral sciences are alike in that they attempt to explain regularities in the universe. Scientists from each of these broad disciplines develop theories to capture and summarize reality, to make predictions, and to guide future research. They share research strategies, including controlled observation, experimentation, and the statistical analysis of results. Each behavioral science provides a foundation for practitioners. Even as there are engineers and physicians who solve real problems for real people living in a real world, there are clinical and applied psychologists, applied anthropologists, clinical sociologists, and political scientists who attempt to apply their knowledge for the benefit of humankind.

Of course, there are also some differences between the physical, biological, and behavioral sciences. Sometimes in the behavioral sciences controlled observation is difficult or impossible, and when quantitative data *are* available, the level of precision achieved by physical scientists cannot be attained by behavioral scientists. Yet differences in focus and perhaps current levels of success should not cause us to overlook similarities in assumptions, outlook, and methods.[12]

Different behavioral sciences bring different strengths to the SETI enterprise. Psychology sensitizes us to the broad diversity of animal and human behavior. Its exploration of human emotions and thoughts helps us understand people's attitudes toward extraterrestrial intelligence and toward the search. Psychology helps us to predict individual responses to contact and to understand the many lines of defense against hysteria or panic under stress.

Anthropology, despite its literal definition as "the study of man," is eminently suited to play a conspicuous role in the search for extraterrestrial intelligence. Anthropologists have a keen appreciation of differences between cultures and may find it relatively easy to relate to extraterrestrial societies. Their understanding of ancient and modern languages may help us decode extraterrestrial transmissions and establish a basis for responding. Furthermore, anthropologists are some of our best forensic social scientists. From minimal traces, the "stones and bones" experts, archaeologists, can reconstruct entire civilizations. These skills could prove useful under certain conditions. For example, archaeologists might be able to reconstruct a long-dead but highly advanced civilization, thus allowing lessons from their past to propel us into the future.[13]

Some of the most interesting issues have to do not with

individuals but with organizations, societies, and interacting societies, or "interstate systems." Consequently, we can learn much from sociology, with its emphasis on social organization and comparative analyses of large social entities, and from political science, with its understanding of power and the relations between nation-states. These fields can help us make an educated guess as to whether extraterrestrial societies are likely to be belligerent or peaceful. As we shall see, an understanding of alliances, blocs, and coalitions may also explain why, if there are many intelligent civilizations out there, we have yet to hear from even one of them.

Behavioral scientists underscore cross-species, cross-cultural, and situational variability in behavior, thereby reducing the dangers associated with egocentric perspectives. Whereas few if any of us consider ourselves the center of the universe, some of us sometimes act that way and many of us are given to overrating our significance in the larger scheme of things. Symptoms of egocentrism include a firm conviction that one's own ways are right and proper, and that one's interests should be placed above those of everybody else. Personal beliefs, no matter how idiosyncratic, provide the framework for understanding and evaluating other people.

Groups, as well as individuals, develop self-centered ways of thinking and apply their standards to other groups. This is true of groups of all sizes, ranging from very small ones, such as dating couples or families, to very large ones, such as nations. Referring to group-centered views, social psychologists refer to "own-group bias," sociologists refer to "in-groups and out-groups," and anthropologists refer to "ethnocentrism." Such centering can proceed at still higher levels: "anthropocentrism" means relating everything to the ways of humankind.

Historically, such biases may have been useful. A shared set of values regarding who we are, what we are doing, and what is "right" makes it easy for people within our group to get along with one another. Furthermore, a lack of empathy for outsiders may lessen the pangs of guilt when someone else's group must lose in order for our group to win. However, in this era and in future ones, competitive strategies are less promising. Advanced technology and the promise of abundant wealth could allow us to redefine win–lose situations as win–win situations in which everyone can prosper.

Excessive reverence for our own ways may hinder the search for extraterrestrial intelligence if it limits our thinking to a relatively narrow range of possible life-forms or restricts our search to those forms that are most similar to us.[14] Also, an assumed superiority of oneself, one's group, or one's species promotes a judgmental approach, which in turn sets the stage for miscommunication and conflict. As Ian Ridpath has pointed out, we need professionals who can "explain our culture to the visitors, and discover more about the behavior and ethics of the aliens so that neither group offends the other by some unintended rudeness or sacrilege."[15]

Guessing the Number of Extraterrestrial Civilizations

The likelihood that SETI will lead to the discovery of extraterrestrial intelligence depends in large measure on the sheer number of extraterrestrial civilizations that exist simultaneously with our own. The more civilizations "out there" right now, the more likely we are to find one of them. Attempts to grapple with the question, "How many?" begin by identifying the physical, biological, and social conditions necessary for such civilizations to arise and persevere, and then making educated guesses as to the prevalence of such conditions in our own Milky Way galaxy (or, if one prefers the grand perspective, the universe). How many stars are there? How many stars have planets? On what proportion of these planets will life evolve, and what proportion of these life-forms will qualify as intelligent? This line of questioning, systematized by astronomer Frank Drake in preparation for a meeting at National Radio Astronomy Observatory at Greenbank, West Virginia in 1961, sets forth the sequence of questions that must be asked.[16]

The process is reminiscent of a decision-making procedure known as "elimination by aspects." In this procedure, one begins with a large pool of candidates. Then, by successively applying different criteria, unsuccessful candidates are eliminated. Normally, the process is repeated until only one alternative remains. Amos Tversky observed this long ago in a television commercial for a vocational training program[17]:

> There are more than two dozen companies in the San Francisco area which offer training in computer programming. The announcer puts some two dozen eggs and one walnut on the table to represent the alternatives, and continues: "Let us examine the facts. How many of these schools have on-line computer facilities for training?" The announcer removes several eggs. "How many of these schools have placement services that would help you find a job?" The announcer removes some more eggs. "How many of these schools are approved for veteran's benefits?" This continues until the walnut alone remains. The announcer cracks the nutshell, which reveals the name of the company, and concludes, "That is all you need to know, in a nutshell."*

To apply Drake's procedure, we begin by estimating all of the stars that could host intelligent civilizations and then eliminate successive groups of candidates on the basis of educated guesswork. But, unlike the television announcer, SETI enthusiasts seek to retain rather than eliminate as many alternatives as possible. That is, they hope that when the winnowing is completed, a generous pool of advanced extraterrestrial civilizations will remain. After all, if the pool is skimpy, then the chances of finding even one extraterrestrial civilization are uncomfortably close to zero.

Stars

Stars have been defined as "celestial bodies consisting of large, self-luminous masses of hot gas held together by their own gravity."[18] Stars gravitate into loose, symmetrical disk-shaped forms called galaxies. There are perhaps 100 billion galaxies in the universe, each containing hundreds of billions of stars. Carl Sagan and Iosev Shklovskii suggest that there are at least 10^{20} stars in the universe—that's 1 followed by 20 zeros.[19] The NASA–SETI panel stated that, given 1974 population figures, "in the enormous emptiness of space we can now recognize so many stars that we could

*Amos Tversky, "Elimination by Aspects: A Theory of Choice," *Psychological Review*, **79**, 297 (1972). Copyright 1972, the American Psychological Association, reprinted with permission.

count one hundred billion of them for each human being."[20] New stars are forming all the time. Thus, we are off to a great start.

However, not all stars are promising hosts for life. Some are dual stars that orbit around each other, and it is unlikely that their planets could achieve stable orbits. Stars can be too young, too old, too hot, too cold, too large, or too small to host life-sustaining planetary systems. Very new stars lack potential because they have not been around long enough to support the evolution of life. Very old stars may have already run their course, with the result that any life that once existed within their planetary systems is now extinct. Thus, many stars are, for one reason or another, not congenial to life. Yet, the pool is by no means small, because G stars, K stars, and perhaps M stars are suitable for life. Together, this group, which excludes very hot blue stars, comprises stars ranging from solar yellow to cool red.

Planets

The conventional view is that planets, not suns or space itself, are the necessary sites for the origination (but not necessarily the perpetuation) of life. If planets are relatively common companions of stars, life would have many chances to form outside our solar system, but if planets are rare, it would have few opportunities to do so. For the past 50 years, scientists have overwhelmingly favored the nebular hypothesis, which implies that the formation of planets is a common event. According to this hypothesis, slowly rotating clouds of gas and dust eventually collapse into flattened disks. After they settle into thin sheets, particles draw together to form solid objects called planetesimals. These, in turn, attract additional matter within their gravitational fields and form larger bodies. Over the aeons, suns and planets form at the same time.[21]

Until recently, apart from one planet that accompanies a special kind of (lethal) star known as a pulsar,[22] there was no firm evidence that planets existed outside our solar system.[23] Then, in early 1996 came confirmation of two huge planets orbiting sunlike stars many light-years away.[24] If planets are common, why have they been so difficult to find?

It is not that astronomers have not looked, and it is not

necessarily because there are few planets outside our solar system. The problem is that distant planets are very difficult to detect: Pluto, as far as we know the most distant planet in our own solar system, was not discovered until 1930. Stars are relatively easy to find because of *their* size and brilliance; planets are almost impossible to find because of their size and brilliance. Stars outshine planets by as much as a billion times. Astronomers are forced to rely on indirect techniques, that is, looking for the effects of planets on their stars in the form of stellar "wobbles." However, such techniques require very precise measurements, in some instances on the order of measuring a dime at 300 miles. Once the first extrasolar planet was detected, others followed in short order: by October 1996, planets had been identified around six stars and there was a tentative identification of a planet orbiting a seventh.[25] The pace will accelerate as NASA's new space telescopes come online. The goal of NASA's Origins Program is to locate and photograph Earthlike planets outside our solar system.

We must, of course, disregard those planets that cannot support life. While early science fiction writers found almost limitless niches for the evolution of life, SETI scientists tend to focus on Earthlike planets that "have a stable source of energy ... and plenty of raw materials in usable forms."[26] For example, planets that are too close to their suns will have no water or only boiling water, and those that are too far away will have only frozen water; neither steam nor ice seem conducive to the evolution of life. A small planet is unlikely to hold an atmosphere; a large planet may have a very dense, compressed atmosphere, something like a sea of molasses.

If life does gain a toehold, it can increase a planet's suitability for more complex forms of life. A by-product of the simple plant life that dominated Earth hundreds of millions of years ago was oxygen, which made Earth habitable for oxygen-breathing animals. Our oxygen atmosphere was both an effect and a cause of the diversification of living things. Thus far, the result for Earth has been somewhere between 5 and 50 billion species, approximately 40 million of which live today.[27] The point remains that a substantial proportion of planets must be really lousy bets for life: they are arid wastelands, glowing embers, or balls of ice.[28]

The planets discovered in the late 1990s are huge planets whose size makes them relatively easy to detect but also, we suspect, unlikely sites for life. There may be countless planets that have masses comparable to Earth's, but with our current technology they remain beyond our reach. Tobias Owen raises the interesting possibility that giant planets could themselves have Earth-like planets in orbit, providing a "new set of environments on which life might arise and flourish."[29]

Natural endowment is not the only way that a planet could support intelligent life. An alternative route is for immigrants to make that planet suitable. Less than 100 years after our first heavier-than-air flight, we have been to the moon and know how to get to Mars. Furthermore, we have the knowledge and can develop the technology to terraform Mars, that is, make it more Earthlike.[30] This includes landscaping, thickening the atmosphere, adjusting the temperature, and increasing the availability of water. According to one recent analysis, four fusion warheads, dropped on Mars in such a way as to penetrate the surface, would kick up enough dust to initiate a runaway greenhouse effect. This would in turn produce an atmosphere so thick that liquid water would exist, people could walk around with respirators rather than full space suits, and plants could grow.[31] Advanced spacefaring civilizations may have restructured planets in their own solar system or elsewhere to make them inhabitable. For each advanced spacefaring civilization there may be dozens, scores, or even hundreds of planets that are adapted for life. Thus, whereas life may have originated in very few places, it may exist in many places—in huge orbiting colonies, terraformed planets, and immense, self-sustaining "worldships" that transport thousands or millions of passengers on interstellar journeys.

Life

There are at least three ways that life could appear on a planet. First, highly evolved, technologically advanced aliens could move from place to place, establishing colonies on naturally suitable or terraformed planets. Second, life could have originated on one planet and the seeds of life could have landed on other planets (perhaps even ours), there to be nurtured and to grow. Third, life may have evolved independently at many times and in many places in the universe. From the viewpoint of SETI scientists, solar and planetary evolution give rise to chemical evolution, and chemical evolution gives rise to biological evolution, as sure as night follows day.[32]

The evidence is circumstantial but compelling. Spectrographic analyses and other studies suggest that the chemicals necessary for life are widely distributed throughout the universe. These include not only the same elements that constitute living matter, but also the heavy hydrocarbons and amino acids that are among life's more complex building blocks.[33] Organic molecules have been identified in at least 50 galaxies.[34] Life, or at least its biotic precursors, has been produced in laboratory settings. The original experimental work by Stanley L. Miller, Harold Urey, Lesley E. Orgel and others consisted of combining chemicals that were believed to be crucial for life, energizing the primordial soup with electricity, and assessing the results.[35] These studies suggest that even under the tight time frames of human experiments, the biotic precursors of life are fairly easy to produce. Since the necessary ingredients are widespread throughout the universe and the universe has plenty of time, the expectation is that life will evolve again and again.

Intelligence

What percentage of hospitable planets have life-forms that have evolved a suitable level of intelligence? For SETI's operational purposes, a life-form is intelligent if it can make itself known to us, in almost all probability through microwave transmissions. Since most radio telescope searches involve passively listening for radio signals, ET must have radio technology that is at least the equal of what ours was in the 1940s.

One argument in favor of discarding many contenders at this juncture is that a high level of intelligence and technological sophistication do not necessarily confer a survival advantage; hence, we should not expect intelligence to evolve. In some cases, survival may depend on strong and relentless reproductive urges or a high level of aggressiveness. Many species on Earth, for example, have remained basically unchanged for tens or even hundreds of millions of years. Insects, reptiles, and amphibians (among others) have done just fine in the absence of poetry, medicine, organized welfare, high school diplomas, and TV talk shows. But in other cases, as we shall see, survival depends on intelligence and investing tremendous care on a limited number of offspring.

By our operational definition, Earth has one intelligent species, humans. The crucial question is whether our evolution depended on a remarkable series of coincidences, a series that could almost never be duplicated, or whether, in the course of evolution, there were many paths to the same result. If an uncanny chain of coincidences was required to produce a species such as *Homo sapiens*, humans should not expect many intellectual peers or superiors. If, however, there is some play in the system, if there are many routes to the same general destination, then we might expect a widespread distribution of intelligence. It is here that the battle lines are the most clearly drawn, with some scientists arguing that the evolution of intelligence is a rare and unlikely event and others finding many evolutionary paths.

Paleontologist David Raup has both bad news and good news for those who hope that intelligence has evolved elsewhere. SETI enthusiasts frequently envision biological systems as passing through a series of stages in which they gain intelligence, technology, and radio communication. Raup does not find, within the paleontological record, evidence suggesting the predictability, never mind the inevitability, of this sequence. The good news is that intelligence is not necessarily restricted to a limited range of life-forms. Despite his skepticism that evolution inevitably leads to a technological civilization, Raup is a strong SETI supporter. "Only by discovering biological systems elsewhere in space," he writes, "will we really have the means to know whether our own biological system has predictable patterns not yet recognized."[36]

Paul Davies suggests that the paleontological record may not be as much of a jumble as some paleontologists think.[37] We are discovering that, in many areas, what appears to be chaos is actually order. From Davies's perspective, the probability of life's originating through the random arrangement and rearrangement of molecules is so low that we should be surprised that life arose even once. (Davies estimates the odds against random permutations of molecules assembling DNA to be about $10^{40,000}$ to 1.)[38] He is among those who believe that the self-organizing processes that are abundant in physics and chemistry provide a possible key to the origin of life and consciousness. As Stuart Kauffman points out, order did not result from the "grinding away of a molecular Rube Goldberg machine, slapped together piece by piece by evolution"; rather, order is "spontaneous, a natural expression of the stunning self-organization that abounds in very complex, regulatory networks."[39] It is not necessary for chains of 2,000 enzymes to assemble one by one; in effect, complex mixes of chemicals spontaneously crystallize into life. Thus, Davies adds, the "remarkably ingenious" laws of physics "permit matter to self-organize to the point where consciousness emerges in the cosmos."[40] The view that life is a natural property of complex chemical systems means that the correct motto for life is not "We the improbable" but "We the expected."[41]

As Dean Falk reports, species that lack a common ancestry but that must adapt to similar ecological conditions develop similar characteristics, including neuroanatomical features.[42] For example, larger body sizes are associated with larger brain sizes in carnivores, ungulates, sharks, and birds, even though they evolved independently. There is limited evidence that as the size of the brain and the number of neurons increase, cortical specialization occurs. Additionally, as one of the brain's functional areas gains in size and complexity, so do other areas, to handle the first area's output. Falk's point is that although evolution is unlikely to lead to identical species, similar features have evolved in both living and extinct mammals. The reason that only one intelligent species of hominids remains on Earth, suggests Falk, is that we eliminated our closest terrestrial competitors long ago.

Comparative psychologist Lori Marino also finds a certain inevitability in the evolution of intelligence. She notes that to our knowledge, trends toward increased relative brain size and increased intelligence have never been reversed. She proposes that even though neurons are biologically "expensive," it may be easier to develop greater intelligence than new physical adaptations. Relative to, say, developing anatomical and metabolic characteristics that would have allowed us to meet our current speed records on land, in the sea, and in the air, it was easier to evolve brains that allow us to build speedboats, racing cars, airplanes, and rockets! If, in fact, adding information-processing capacity is the most feasible way of relaxing the restrictions imposed by the physical environment, then "increasing the amount, complexity and speed of information processing [intelligence] may be the direction towards which all organisms move."[43]

Longevity

It is possible that we might somehow stumble upon messages or relics from a long-dead civilization, but for practical purposes, all that counts in SETI are civilizations that exist concurrently with our own. Thus, the length of time that a civilization can endure is critical. If highly technological civilizations last but a brief period of time, then one would expect very few concurrent with our own. If, on the other hand, highly technological civilizations endure indefinitely, we might expect many to overlap. *Homo erectus*, a primitive but nonetheless technologically aware species extended over a million years, but then, *Homo erectus* was not electromagnetically active.[44]

Speculation on the longevity of civilizations has been shaped by war and by our inability to maintain our planet's health. Would other civilizations succumb to nuclear warfare and bring about their own demise a few centuries or millennia into their high-technology era, or would they find a means to prevent such destruction? Estimates made in the 1960s, a period marked by a hot war in Vietnam and a cold war elsewhere, tended to be pessimistic: many civilizations would blow themselves up early on, it was thought, but perhaps the few civilizations that survived would continue for millions or even billions of years.

The thawing of the cold war, stemming in part from the dissolution of the Soviet Union, has generated greater optimism: after all, if we got through 40 years of dancing on the edge of nuclear disaster, perhaps other civilizations have done the same. Frank Drake notes that Iosef Shklovskii, initially a great proponent of the search for extraterrestrial life and coauthor with Carl Sagan of the seminal work *Intelligent Life in the Universe*,[45] later became pessimistic about the search because it seemed unlikely that the world could avoid a nuclear holocaust. Thus, notes Drake, Shklovskii's pessimism was based on a "political rather than mathematical calculation." Had Shklovskii lived to our period of glasnost, Drake believes, "I am sure he would have changed his mind again, and signaled, with his characteristic smile and shake of the head, his renewed enthusiasm for SETI."[46]

Of course, as Joseph Baugher notes, nuclear war is only one risk; a society can fall in other ways.[47] One possibility is that rapid population growth will deplete essential resources. Human societies survive, in part, because of the ready availability of cheap fossil fuels; we will need to find new energy sources as these fuels are exhausted. Another risk is environmental pollution, a by-product of energy production and manufacturing. Still another risk is the genetic deterioration that could result from mutations or from conditions that allow not only fit organisms to survive but unfit ones, too.

Beyond those factors identified by Baugher are cataclysms that could jeopardize life on all the planets within a solar system. On Earth, mass extinctions occur once every 26 million years on average. One theory is that Earth has an invisible companion star (called Nemesis) that disturbs the orbits of comets, causing them to shower Earth and create the devastating equivalent of a nuclear winter.[48]

Attempts to stabilize population growth, restrain the use of energy, limit technology, and care for the environment may allow some societies to endure, but at the price of forgoing contact with other worlds. A society where "small is beautiful" may survive, but it is unlikely to devote the necessary energy to seek other civilizations. Thus, suggests Baugher, the society that we will hear from will not only have overcome the kinds of problems that confront us on Earth but will also have solved them without undermining their technological advances. Large and powerful societies, for instance, may be able to fend off an approaching comet or withstand the devastation of one planet if they inhabit several planets or locations within their solar system. Or, they may assure their long-term survival through a process such as cryoconservation.*

*This is the preservation of genetic material through storage in near-liquid nitrogen at temperatures of 77° K. Recently, Michael Mautner[49] has suggested this as a technique for preserving the genetic material of endangered species (including many racial or ethnic groups). To avoid the "instability of human institutions," he recommends locating this material in such places as permanently shaded lunar craters, on the moons of Saturn, or in storage satellites. Such remote and frigid sites are immune to most disruptions and offer the advantage of perpetual refrigeration.

The expectation of SETI scientists is that when we do hear from another civilization it is likely to be older and smarter than ours. Let us assume that we can attach intelligence scores to civilizations as well as to individuals. We will further assume that intelligence is normally distributed throughout our galaxy; in other words, that there is a bell curve. By our standards, some civilizations are rather dull or backward, some are normal (like us), and some have advanced far beyond us. From this bell-shaped distribution we have sampled one case: contemporary terrestrial civilization. We do not know where we fall relative to the rest of the distribution.

If we fall near the top of the distribution, we would have few intellectual peers and might as well quit the search. But what are the statistical odds of this result? A normal distribution has relatively few extreme cases. Most civilizations are bunched in the middle. It could be that we come from the top of the distribution, but this does not seem likely. Drawing one case (human civilization) from the top of the distribution would be like choosing one person for a study of human intelligence and happening to pick Einstein or Mozart. Indeed, one might argue that because of our youth and inexperience* we are one of the *least* advanced civilizations and fall at the bottom of the curve.

According to the assumption of mediocrity, the level of civilization and technological sophistication on Earth should be about average: roughly half of the civilizations elsewhere should be behind us, whereas the remainder should be further along the developmental path.[51] Earth is neither advantaged nor disadvantaged relative to other habitable planets in the race for life. However, since we have just entered an age where we can communicate with other solar systems, we will never hear from those that are behind us developmentally, only those that are at or beyond our level. Furthermore, if we assume that civilizations that work through the minefields of war and environmental disasters last hundreds of thousands or millions of years, then it is these truly ancient (and most likely superadvanced) civilizations that are the most likely to be with us as we enter the 21st century. The

*Human life has been present for about one-hundredth of Earth's geologic lifetime, civilization of one sort or another for about one-millionth of Earth's lifetime, and technological civilization for about one-billionth of Earth's lifetime.[50]

assumption of mediocrity is critical to understanding whom we are likely to hear from, and we shall refer to this assumption throughout this book.

OPTIMISTS AND PESSIMISTS

Frank Drake acknowledges that we know almost none of the values that we need to apply his process of elimination.[52] "Drake's equation," states Emmanuel Davoust, "does not furnish any valid quantitative estimate of the probable number of extraterrestrial civilizations in the galaxy. Rather, it shows us the difficulties and unknowns that still need to be cleared up before serious demographic work can be undertaken."[53] Such cautions have not kept everyone from trying. For example, J. Freeman and L. Lampton have estimated that 250 million communicative civilizations have evolved in the history of our galaxy, and that at present at least 3 million planets host at least apelike intelligence.[54] Carl Sagan estimated that there are "one million civilizations more advanced than ours in our galaxy."[55] Drake and Dava Sobel guess that there are approximately 10,000 advanced extraterrestrial civilizations in the Milky Way.[56] Frank Tipler, on the other hand, argues that the universe must contain 10^{20} stars in order to contain a single intelligent species, and that we should not therefore be surprised if it contains only one.[57]

John Baird explains how to recognize the highly subjective nature of such estimates.[58] When we operate with "hard" observational data we are precise and any digit up to 9 has a reasonable probability of appearing in an answer. However, when we guess, we tend to pick convenient numbers, specifically, 5, 1, and 0. For example, if we have real data on the "fairness" of a coin toss, we might state that the probability of heads is 0.538, but if we are only guessing, we are likely to set the probability at 0.500. Convenient, round numbers appear in many discussions of the Drake equation.

No matter how objective we try to be, our understanding of the world is affected by our wishes, needs, motives, and hopes.

Much of what we see and believe to be external to ourselves is shaped by the inner workings of our minds. Sagan and Shklovskii's pioneering work, *Intelligent Life in the Universe*, recognizes the subjective element in human cognition in a chapter titled "Extraterrestrial Life as a Psychological Projective Test." Their point is that some people have very strong expectations regarding the likelihood and nature of extraterrestrials, even though we do not know if they exist, or, if they do, what they will "really" be like. Because there is "no unambiguous evidence for even simple life-forms," they note, "we are at the mercy of our prejudices."[59] Thus, psychological factors, both unconscious and conscious, affect people's views of extraterrestrial life.

Over the years, psychologists have found many bases for our attitudes.[60] Some are simple and straightforward. We tend, for example, to develop favorable attitudes toward ideas, activities, and people that bring us rewards or somehow better our lives, and unfavorable attitudes toward ideas, activities, and people that frustrate us or make us worse off. Optimists, who believe that intelligent life is abundant in the universe, might delight in appearing on talk shows and revel in favorable attention from the media. Pessimists, who believe that intelligent life is statistically rare, may express annoyance at the diversion of research funds from their own "serious" work and the loss of radio telescope time. Optimists enjoy the accolades of the public and the widespread sales of their books. Pessimists are secure in the approval of a cautious scientific establishment. The influence of social approval on expressed attitudes may be reflected in Drake's recognition that his seniority as a professor gives him the freedom to express views that could incur penalties if he did not outrank his professorial colleagues.[61] As a junior faculty member trying to build points with a conservative establishment he might not be quite so optimistic.

Other bases for our attitudes are our values and ego defenses. Our *values* are our deepest convictions and highest priorities—for example, religious convictions or a strong sense of nationalism. Our *ego defenses* are unconscious processes that make us feel better about ourselves or the world in which we live. Ego defenses protect us from troublesome memories, perceptions, or realizations. They impart a sense of security ("Don't worry; if they are hostile, there is no way their starfleet could get here") or make us feel superior (*Homo sapiens* is the only intelligent species in the galaxy"). Ego defensiveness is one basis for prejudice: sometimes we try to give ourselves a sense of superiority by finding fault with others.

Both values and ego defenses may contribute to the "anti-Copernican conceit," as described by Sagan.[62] Prior to Copernicus, it was widely held that Earth was the center of the universe and that the sun revolved around it. Copernicus expounded the view that in fact the Earth revolves around the sun. This was followed by a succession of discoveries that reinforced the view that Earth was far from the center of things: the discovery that our sun occupies a peripheral location in the Milky Way, the discovery that our galaxy is not in the center of the universe, acknowledgment of billions of other galaxies, and so forth. In the late 19th century, evolution—with its view that all living things are connected—challenged the position that humans are separate from, and superior to, all other beings.

The "anti-Copernican conceit" refers to an exaggerated sense of human significance in the universe, a sense that we are unique and alone. There are both value-expressive and ego-defensive elements here, to the extent that such views are based on religious convictions or have the practical effect of reaffirming our sense of superiority. Thus, a survey by William Bainbridge found that in comparison with people who held other religious views, evangelical Protestants (typified by those in the "born again" movement) were less likely to believe that intelligent extraterrestrials exist and less likely to support SETI. Whereas some respondents had an almost religious fervor for the space program, including SETI, others were threatened by possibilities that could undermine their fundamentalist doctrines.[63]

JUSTIFYING THE SEARCH

Almost everyone concedes that a successful search for extraterrestrial intelligence would be one of the momentous events of all time and would change humanity forever. However, even optimists have to admit that the search could be unsuccessful, at least in the sense that it fails to establish contact with extraterrestrials. Consequently, it is unwise to attempt to justify the search solely on the basis of benefits that might accrue from discovering ET. It is useful, then, to distinguish between two sets of benefits. Outcome-independent benefits are those accruing from the theory, research, and technology that are the foundations for the search itself. These benefits, which do not depend on actually detecting extraterrestrials, include an improved self-understanding, an enhanced knowledge of the universe, technological spin-offs, a useful venue for science education, and better international relations. Outcome-dependent benefits are those that could occur if an extraterrestrial civilization is found.

Improved Self-Understanding

Improved self-understanding is a common theme in defending the search. We expect insight to flow from disciplined speculation on the range of conditions and creatures that might be encountered as we explore the universe with radio telescopes and by other means. As Michael Ruse points out, "Exploring the possibility of life elsewhere in the universe ... puts a bright light on our own powers and limitations [and] forces us to think about ourselves from a novel perspective."[64]

Frank White believes that both space exploration and the search for extraterrestrial intelligence sometimes produce almost transcendental, self-actualizing experiences that radically change people's perceptions of the universe and their own roles within it.[65] SETI, notes Allen Tough, enlarges our views of ourselves and enhances our sense of meaning and purpose. As we learn about cosmic evolution and SETI activities we gain a deeper sense of ourselves as citizens of the universe.[66]

Advances in Science and Technology

Proponents of SETI note that in conducting the search, we advance science and technology. Outcome-independent benefits include an improved understanding of astronomy and other physical sciences, but the potential benefits extend into the biological and social sciences as well. Potential beneficiaries of SETI technology include electrical engineering, computer science, and radio astronomy, with practical spin-offs in such areas as air traffic control. Since the search involves picking out intelligible radio transmissions from other solar systems, it has required massive advances in multichannel radio reception (to monitor millions of frequencies or channels at once) and information-processing systems (to separate intelligent signals from mere noise).

Of course, communicative extraterrestrials might give us shortcuts to new heights in science and technology. Learning about intelligent life elsewhere could spur rapid advances in the biological, social, and behavioral sciences. Consider, for example, the benefits for my own field of psychology.[67] Since its inception, psychology has involved comparisons of groups, not only experimental treatment groups but also naturally occurring groups defined by gender, culture, and species. Though useful, comparisons between humans and other animals are limited by the gaps in intellectual abilities, especially communication abilities. If we were able to study a species of equal or higher intelligence, we would gain a new perspective on human psychology as well as an initial understanding of extraterrestrial psychology.

If contact is made with an advanced civilization that chooses to share their knowledge with us, we might have rapid gains in all areas, including philosophy, science, technology, and the arts. In effect, benevolent aliens could provide us with the solutions to our problems or, perhaps better yet, with the know-how to solve our own problems. A clean, healthy environment, the abolition of major illnesses such as cancer and heart disease, and the alleviation of social blight, including poverty and crime, are conceivable results. Suddenly, we could have information that would help us address age-old questions. Different religions, notes Jacques Vallee, posit different existential models that make contrasting claims about the origin and purpose of the universe. Knowledge of life elsewhere and an understanding of extraterrestrial histories, religions, and cultures would shed light on the prevalence and perhaps even viability of terran existential views.[68] Of course, all such gains would be dependent on the nature and openness of the civilizations that we might encounter and our ability to apply their insights to our society without untoward results (see chapter 11).

Educational Benefits

For years there has been widespread public interest in the possibility of extraterrestrial life, and the level of interest has attained new heights as we approach the end of the 20th century. Because people are interested in SETI, it is a useful vehicle for enhancing the public's knowledge of astronomy, biology, communication, information-processing technology, and related fields. For this reason, public education has been a major program of the SETI Institute, in Palo Alto, California.[69] Adult education includes public lectures and seminar series, workshops for scientists and teachers, special planetarium shows, video programs, portable displays, brochures, and the like. Some SETI enthusiasts have developed college courses, including courses at Northern Arizona University and California's Cabrillo College, that stress the anthropological, psychological, and social aspects of the search.

SETI also provides an entree for involving children in science. The Smithsonian Institution has expressed interest in this topic, and in 1991 NASA joined with the National Science Foundation in a three-year project to develop science activities for children in the fourth through ninth grades. Through these efforts as well as independent ones, SETI has contributed to educational programs in California, Maryland, New York, and elsewhere.

The University of Western Sydney has established a SETI Australia Centre to disseminate information and build support for SETI. According to Ragbir Bhathal, recent years have been marked by a 20 percent decrease in the number of students who are interested in studying science and technology: an alarming trend that could leave Australia poorly positioned to compete in high-technology industries. The SETI Australia Centre has used SETI to develop a curriculum that involves students in the physical and biological sciences through such topics as the origin of the universe, stellar and planetary evolution, the evolution of life, radio communication, and so forth. Whether or not SETI actually detects signs of intelligent life elsewhere, it generates involvement, and the hope is that students will come away with interests and competencies in science and technology.[70]

An informed citizenry puts heavy demands on science fiction writers. In the early days of science fiction, it was possible to build stories around aliens that differed from humans in minor ways or, alternatively, radically different aliens that lived in wild, topsy-turvy worlds that defied the laws of science. Today, critical and commercial success demands aliens that differ from us in significant ways but remain plausible to a highly sophisticated group of readers. To achieve this, contemporary writers engage in "world building."[71] This begins with choosing the correct kind of star, adding a believable planet and then endowing it with creatures and cultures that make evolutionary sense. In a sense, world building consists of successfully guiding a potential civilization through the various junctures set forth in the Drake equation. In developing an imaginary civilization the world builder has to apply real-universe astronomy, biology, and behavioral science, or risk the wrath of the educated reader.

Improved International Relations

During the 1950s, social psychologist Muzafer Sherif and his associates studied intergroup relations among small groups of preadolescent boys.[72] After two groups—the "Rattlers" and the "Eagles"—were formed in a summer camp, these investigators set up a competition in which only one group could win. This resulted in the symptoms we associate with intergroup conflict: initial good sportsmanship disappeared. Each group began to overrate itself ("We are good") and underrate the other ("They are bad") and to prepare for physical confrontation by stockpiling hard apples for "ammunition".

Establishing intergroup conflict was relatively easy; eliminating it was difficult. The researchers tried many methods before they found one that worked: confronting the boys with problems that could be surmounted only by the concerted action of both groups. For example, all the boys had to work together to push a chuck wagon out of a rut to retrieve the food supply for a barbecue.

The symptoms of conflict between the Eagles and Rattlers bore a clear if imperfect resemblance to the symptoms of international conflict. In consequence, the strategy of reducing conflict through the imposition of superordinate goals arises in discussions of many types of intergroup conflicts, including international relations. *Superordinate goals* are those that have overriding importance to two or more groups and that require intergroup cooperation to attain. They are captured by the wish "If only we could get people to pull together...." Space exploration is commonly proposed as a superordinate goal to help reduce the differences between the United States and other nations. Specific proposals include a manned mission to Mars and the SETI program. The search for extraterrestrial intelligence is a natural for international involvement. To cover the entire sky by means of radio telescope, it is necessary to survey the sky from both the Northern and Southern Hemispheres. Furthermore, as the world rotates on its axis, the source of the electromagnetic radiation could be lost from any individual telescope's view. As it dips below the horizon, continuous tracking will depend on its being picked up by a radio telescope in another country. Finally, as we shall see, most experts suggest that once extraterrestrial intelligence is detected, the different nations of Earth must work together to plan a unified response.

CONCLUSION

Since the earliest of times, humans have speculated on the existence of life on other worlds. Throughout history, the prevalence and strength of beliefs in extraterrestrial intelligence have varied as a function of the state of theology and science.[73] Developments over the past half century have led many contemporary scientists to conclude that life may be fairly common in the universe and that we may detect evidence of the more advanced forms of this life through a radio telescope search. Is this imagination or is this science? Ronald D. Brown captures the view of many SETI proponents when he says that SETI involves models that are rational, internally consistent, and capable of being tested at various points by observation and experiment.[74]

Today, the search for extraterrestrial intelligence is a multidisciplinary effort involving many fields. SETI, notes anthropologist Ben Finney, provides a meeting ground where scholars from all disciplines, including the social sciences and humanities, come together to discuss common interests. Not all of the participants agree with one another, but they do meet and talk. In effect, SETI offers a bridge between very different academic cultures: physical sciences, biological sciences, behavioral sciences, and the humanities.[75]

Early on, NASA panels identified several roles for behavioral scientists who might like to contribute to SETI. Today, such roles include conducting comparative and cross-cultural studies intended to build our knowledge of communication and behavioral diversity, improving our search procedures, and perhaps one day communicating or even negotiating with newfound acquaintances. Behavioral scientists can also contribute to SETI by forecasting, monitoring, and perhaps even influencing human reactions to various search outcomes. The psychology of the search process, the psychological and social aspects of extraterrestrial biosocial entities, and the consequences of discovery for people, institutions, and cultures are the focal points for the remainder of this book. My premise is that assembling pertinent theories and findings is a fruitful intellectual activity even in the absence of a positive search outcome, and that we are better off starting now than waiting until after intelligent extraterrestrial life has been detected.

Of course, not everyone supports a long-term, expensive search for extraterrestrial intelligence. Opponents include people who are pessimistic about the likelihood of achieving contact, people who are pessimistic about the results of contact, and people who have other priorities for the resources that are devoted to the search.

First, whereas one can point to outcome-independent benefits, for many people it is the outcome-dependent benefits that justify the search. Pessimists, who believe that the chances of contact are essentially nil, do not see outcome-independent benefits as justifying the search. Since (according to their view) the search is unlikely to meet with success, why conduct it?

Second, even if contact occurs, the results may be disappointing. The information that is exchanged, particularly at first, may be dull and uninspiring, and if the civilization is more than a few light-years away, it may be difficult to develop anything that resembles a "conversation." Hoping for benevolent technocrats that will share their wisdom is like wishing for help from Santa Claus! Even more pessimistic is the view that we won't like or get along with our newfound neighbors. Under worst-case scenarios we could be wiped out by a plague, subjected to military domination and slavery, forced to participate in painful and humiliating research projects, or even eaten.

Finally, many people think that, given the range and depth of problems that plague humankind, investing in a search for creatures that might or might not exist should receive very low priority. In light of the amount of money that has been spent on physical and life-science research in the past half century, the funds devoted to SETI barely qualify as a drop in the bucket. Nonetheless, there are immediate and compelling needs on Earth. These include medical, psychological, and social problems that could be ameliorated by the money spent on the SETI effort.

SETI proponents admit that the outcome is uncertain and that any success may well come after their lifetime. Some of the pioneering SETI scientists who were interviewed by David Swift note that SETI activities occupied only a small proportion of their professional time.[76] If a particular SETI project involves a continuous long-term involvement, project members take turns so that nobody has to neglect their "mainstream" research. Junior scientists who have yet to establish their careers are discouraged from getting "hooked" on SETI and gently steered toward more conventional fields, which yield higher chances of making discoveries and obtaining publishable results. The scientists who discover extraterrestrial civilization may be remembered for all time. Yet full-time dedication to the search would be like putting all of your money in one slot machine and hoping to get four "genies" in a row. Slot machine jackpots are fun to think about, but we cannot count on them, and we must make our investments accordingly.

TWO

LISTENING

In February 1992 two teams assembled to role-play the first human–extraterrestrial contact. This exercise had been undertaken eight times before by members of the nonprofit organization known as CONTACT, but this particular role-playing was distinguished by extensive and meticulous (well, near-meticulous) planning.

A year earlier CONTACT's board of directors had decided to make this a sophisticated, technically credible exercise to attract prestigious sponsors. To this end, two teams were assembled, one to develop the alien culture and transmission and the other to decode the transmission and plan the human response. The alien team included two noted science fiction writers, a physicist, a geologist, two psychologists, a professional artist, and, on an intermittent basis, several other knowledgeable individuals. Preparation included far-ranging consultations with biologists, astronomers, and other scientists. Artifacts were prepared: a finely sculpted model of the ET (which looked something like a coatrack), a globe, a map, and a computer-animated movie of a flight over alien territory. The human team, including physicists, computer scientists, and some human-factors experts, stood ready to interpret the incoming message. The human team, involving perhaps 16 members, was linked to many consultants via electronic mail.

Finally, after a year of preparation, the moment arrived for the first "transmission." The alien team was to send its message to the human team by computer. All seemed ready. Then, realization dawned. The alien team was equipped with IBM or IBM-compatible computers. The human team was outfitted with Apple Macintoshes. Nobody had remembered to bring the software that makes it possible to link the two systems. Ultimately, the problem was solved courtesy of a nearby university, but the lesson was not lost. If human computers have difficulty communicating with each other, what hardware and software problems might arise when we try to communicate with extraterrestrials?

Assuming that there are other advanced technological civilizations in the galaxy, how might we contact them, or at least confirm their existence? There are many possibilities. Optical telescopes might allow us to detect visual, infrared, or other indicators of their presence. At some future date, human spacefarers may stumble upon extraterrestrials, or in keeping with some flying saucer scenarios, aliens may land on Earth. A careful hunt might lead us to an alien artifact, perhaps a small robot probe somewhere in our solar system or perhaps something left on the moon. Or, it has been proposed (but never proved) that evidence of extraterrestrial visitors might be found among Earth's historical treasures—perhaps in a pyramid, perhaps at a flea market or garage sale. Or, as suggested by SETI scientists, we may someday intercept a microwave transmission that confirms their existence.

In the discussion that follows, it is important to keep in mind the vastness of the universe. Perhaps the best-known unit that scientists use to measure great distances is the light-year, the distance that light travels in one year. Since light travels at approximately 300,000 kilometers (or 186,000 miles) per second, a light year is 9.46×10^{12} kilometers (or 5.897×10^{12} miles). As noted in chapter 1, anything moving at the speed of light would reach the moon in less than 2 seconds and pass our outermost planet, Pluto, in about $5\frac{1}{2}$ hours. The nearest neighboring star, Proxima Centauri, is 4.3 light-years away, and the nearest neighboring galaxy is about 75,000 light-years away. The Andromeda nebula is about 2,300,000 light-years away, and the edge of the visible universe (the surface of the primordial fireball seen by radio telescopes) is about 15 billion light-years away. These enormous distances will make it incredibly difficult to find, let alone interact with, extraterrestrial civilizations.

SPACE TRAVEL

Certainly, the most intriguing possibility is direct physical contact with intelligent extraterrestrial life-forms. This could come about as Earth's own spacefarers move farther and farther into space, or it could result from ET explorers fanning out throughout the galaxy and eventually reaching Earth. This contact scenario is a staple in science fiction literature and was salient in the 1970s, when some of the early social scientific volumes on human–alien contact were published.[1]

At that time, the United States still maintained a vigorous space program: the *Apollo* moon landings were under way, the extended-duration *Skylab* missions were still in progress, and plans were afoot for utopian societies to be created on huge colonies orbiting Earth. Under such conditions it is not surprising that behavioral scientists considered the possibility of humans and extraterrestrials encountering each other.

Fast Ships and Slow Ships

Direct contact requires that extraterrestrials, humans, or both have the means to move from one solar system to another. This means some sort of spacecraft (beings constructed of pure energy might be able to travel at the speed of light, but they would be unable to stop). A "fast ship" is one that can traverse immense distances in brief periods of time. The ideal, perhaps, is the starship *Enterprise*, streaking through the universe in some sort of hyperdrive at approximately 216 times the speed of light. For 20th century humans, at least, time and energy requirements are likely to prevent us from flitting about the galaxy like Captain Kirk and Mr. Spock. These restrictions are imposed by the sheer distances that must be covered and by the laws of physics. Some extraterrestrial civilizations may have been able to overcome these constraints, but to do so effectively would require technology that differs radically from our own and perhaps require drawing on laws of nature that are as yet unknown to us.

There is a trade-off between fuel and speed, such that small gains in speed demand disproportionate expenditures of energy. That is, the faster you want to go, the more fuel you need; the more fuel you need, the heavier the rocket; the heavier the rocket, the slower it will go; and so forth. Lawrence M. Krauss calculates that the fusion engines of the starship *Enterprise* would require 6,561 times the entire ship's mass in fuel to accelerate to half the speed of light, and to stop again. Moreover, the captain would have to allow two and a half months to accelerate and decelerate to keep forces within a range that the crew could withstand.[2] Ronald Bracewell estimates that to achieve 99 percent of the speed of light, the liftoff weight of an interstellar ship would have to be 1 billion times the payload weight.[3] Furthermore, at very high speeds, it would be impossible to avoid contact with space debris, and at exceptional speeds collision with even a small object could lead to the destruction of the spacecraft. An electron that is accelerated to just short of the speed of light would hit with the impact of a Mack truck traveling at normal speed.[4]

In a review of anticipated 21st-century propulsion systems, Robert L. Forward discussed propulsion methods that offer improvements over our current chemical rockets.[5] These include various types of nuclear fission and fusion propulsion, exotic chemicals that produce more energetic reactions than our current fuels, and a number of engine-free and fuel-free devices, such as solar sails propelled by photons from the sun.

There are psychological as well as physical constraints to implementing some of these advanced systems. For example, nuclear rockets may be relatively reliable, cheap, safe, and clean. Yet, notes Forward, it is unlikely that the public can be convinced of this, so despite their safety and efficiency, nuclear propulsion systems are unlikely to be politically viable. Or, a potentially useful propulsion system may be ignored because it has a science fiction ring to it. For example, the *Saturn V* rocket that lifted the

Apollo rockets to the moon had a 622:1 ratio of rocket liftoff mass to payload. An antiproton annihilation propulsion system capable of lifting mass to low Earth orbit would have a ratio of less than 3:1. Furthermore, recent theorizing suggests that antimatter rockets may be much cheaper than we formerly believed.[6] But because "the word *antimatter* still evokes raised eyebrows, mental images of *Star Trek*, and stifled giggles" from executives and managers, states Forward, we are not developing an important propulsion technology that could open up the solar system.

At our present stage of development the biggest headache for rocket scientists is not setting interstellar speed records but finding ways to reduce the costs of lifting material out of Earth's gravity well and propelling it to nearby orbital, lunar, or planetary sites. At a recent meeting on inexpensive ways to get people to Mars, I was struck by the variety and ingenuity of the proposals: big, cheap rockets; lightweight, high-tech, but finicky rockets; rockets that could get by with less fuel because their buoyancy moved them the first few feet upward in a pool of water right before ignition; hybrids that take to the air with standard jet engines and are then loaded with rocket fuel for their journey outward; missions that reduce weight by sending people on lifetime (one-way) missions. The primary focus of 21st-century rocket scientists will be improving our ability to get around in our own solar system, not making interstellar journeys.

In some science fiction stories, spacecraft travel faster than light and immense interstellar distances melt into insignificance, but physics provides scant encouragement for superluminal, or faster-than-light, travel. The theory of relativity imposes a cap at the speed of light; it is not possible to travel or to transmit information beyond this speed. Over the years theorists have suggested certain loopholes in physics such that faster-than-light travel, or at least communication, might be possible. Nick Herbert's book, *Faster Than Light*, gives us a fascinating look at these theories, but what we find are glimpses of hope followed by disillusionment.[7] Thus, in theory, huge, rapidly rotating cylinders might achieve the speed of light, but not only would there be hopeless energy requirements, the cylinders would also collapse on themselves before achieving the necessary rotational speed.

Thus, as far as we can tell right now, both manned and robotic missions will have to proceed below the speed of light, although such missions may come closer to that speed as new technologies are devised. Precisely because of advances in science and technology, Robert Forward suggests that we should not undertake missions unless they can be completed in less than 50 years.[8] If a mission takes longer than this, Forward suggests, the spacecraft is likely to be overtaken by a ship launched decades later but with a better propulsion system. How would you feel if you set out for Alpha Centauri on a spacecraft powered by nuclear bomblets only to be passed decades later by your descendants riding in a ship powered by antimatter?

As Eric Jones and Ben Finney have pointed out, an alternative to the fast ship is the slow ship, or nomad, that requires several generations of spacefarers to reach distant stars.[9] Their view is that to spread throughout the universe, *Homo sapiens* could hop from place to place, eventually jumping from one solar system to another, one at a time. Jones and Finney's book describes how this could be done starting with a relatively small group of people and minimal equipment and supplies.[10] If one is willing to settle for slow progress, the need to achieve very high rates of speed, with the accompanying energy requirements, is set aside.

The model here is the island-hopping migrations of various populations throughout the Pacific. (Finney himself has sailed solo from island to island, taking advantage of normal weather conditions and ocean currents.) Migrations across the Pacific did not occur in one huge jump, but involved successive jumps from location to location. Such an approach might make it possible to colonize distant stars.

Although we ourselves have yet to develop interstellar vehicles, some extraterrestrial societies may have done so, and as their starships move around the heavens, they may emit spectral signatures in the form of gamma rays, X rays, and visible light that reveal their nature. Extrapolating from progress in nautical engineering, Robert Zubrin estimates the acceleration velocity, exhaust velocity, and power requirements of a "typical" starship on the order of a million tons. Zubrin then identifies the characteristics of four propulsion systems that might be used to propel the vehicle: fusion rockets, fission rockets, antimatter rockets, and magnetic sails. The distances at which different types of starships might be detected range from being undetectable at interstellar distances to being identifiable at several thousand light-years, depending on such considerations as the type of propulsion system employed, the orientation of the exhaust nozzle relative to the observer, and the location and sensitivity of the detection equipment.[11]

Loopholes

There are many proposals for easing interstellar travel. These include suspended animation, time dilation, and space-warp technology. These are, in effect, "loopholes" that might allow us to compensate for our inability to achieve speeds near that of light, but even if some of these loopholes can be made to work, we are a long way from being able to capitalize on them.

Temporarily suspending consciousness and bodily functions might be one way to ameliorate the effects of long-term space travel. In science fiction, suspended animation is a convenient gimmick to allow spacefarers to arrive at their destination refreshed by a 100,000-year nap. Unfortunately, this requires substantially more than one or two sleeping pills.

Cryonic suspension might be possible. Through this procedure ambitious spacefarers would be frozen at very low temperatures, to be thawed and reactivated upon arriving at their destination. Indeed, cryogenic suspension is sometimes touted as a way to evade death or at least get a second chance at life. A person who has recently died can be kept in a cryogenic tank to be revived when technology advances to that point where frozen corpses can be revitalized and the cause of the original death (heart disease, cancer, etc.) overcome. Customers may have had either their entire bodies or simply their heads preserved with the intent of rejoining the human scene decades or centuries later. Presumably, a modest initial investment in a certificate of deposit for 500 years or so would not only pay for the revival but also support a life of comfort and ease.

A problem for people who choose cryonic suspension as the road to immortality is that the freezing process must occur very shortly after death. A general rule of the thumb is that a person who is not revived within 9 minutes or so following cardiac arrest is unlikely to be revived without risking profound mental impairment. (This interval may be longer if the person has undergone oxygen starvation coupled with a lowered body temperature.) This problem would not confront spacefarers, who could be frozen while completely healthy and in the prime of life. Unfortunately, there remains the problem that water expands when frozen, which can lead to the rupture of cell walls. The damage may be reduced by very fast freezing followed by very slow thawing. Nevertheless, if such damage occurs in the brain, one might question not only whether mental functions will resume but also whether there will be adequate memory to preserve a sense of continuity, personality, or self. To me, the compelling demonstration will not be the revival of a frozen rabbit or rat; rather, it will be the revival of a trained animal that is still capable of performing a previously learned response.

Einstein's view of the universe includes a fundamental interdependence of distance and time. If an object moves at a terrific speed, time as we know it may either be sped up or slowed down, depending on perspective. This phenomenon is called time dilation, and it does not refer to the mere experience of time, but to all the processes that we associate with time, including respiration, reproductive cycles, and aging.

To observers in space, time will fly back on Earth. On Earth, members of one's cohort will age and die while travelers in space will age at a leisurely pace. It would be possible for spacefarers to set forth on a mission and return only to discover that all traces of their families had disappeared as a result of the natural passage of time: a mission that involved 10 years inside a spacecraft could take 25,000 years back on Earth. The world itself could change so dramatically as to be unrecognizable, and nobody on Earth may remember the mission participants or the purpose of their trip. What time dilation does offer is the opportunity to complete a long-distance mission within the lifetime of the spacefarers.

At some point, notes Paul Halpern, we might be able to find cosmic shortcuts that allow us to jump to any region of the universe.[12] The key is finding topographical quirks in space that interrupt normal space–time continua. Entering one of these areas would make it possible to cover large distances almost instantaneously or to even move backward in time. Some such shortcuts are consistent with Einstein's general and specific theories of relativity and other important theories.

At first glance, black holes hold some promise. These are very tightly compacted matter with such ferocious gravitational forces that not even light can escape from them. Black holes come in many sizes, and it is the larger ones that offer the most intriguing possibilities. Some of these rotate; the skilled spacecraft commander might be able to pick up some of the momentum and use the rotational force like a giant slingshot to move into space at a terrific speed. Really big black holes may affect the curvature of space and time in such a way as to permit jumps from one place or time to another.

However, attempts to use such shortcuts could be fatal. Some explorers might meld with the black hole as airplanes occasionally meld with mountainsides, only seamlessly and with no traces. Other explorers might enter tunnels that collapse once they have been entered. The spacecraft and its occupants won't necessarily be instantly obliterated; instead, they may be pulled into long skinny strings, somewhat reminiscent of pulled taffy. But the biggest problem is that gravitational forces sufficiently strong to keep photons from escaping virtually guarantee that anything that enters a black hole is on a one-way trip.

A possible reprieve is found in the hypothesis that black holes, which absorb matter, are connected to distant white holes, which spew forth matter. Thus it might be possible to make a one-way entrance into a black hole and a one-way exit from a white hole. Yet if white holes exist, they may remain in operation for only a few moments as the huge amount of matter they disgorge collapses in on itself and turns them into black holes.

Another possibility, initially envisioned for Carl Sagan's science fiction novel *Contact* (now a major motion picture) and later developed as an exercise in theoretical physics, is the construction of "cosmic wormholes," that is, specially crafted tunnels that connect different places and times. Wormholes would require incredibly dense matter and would have to be located in remote locations so as not to upset such things as the orbits of planets and moons. At some point it might be possible to devise a "stargate" such as depicted in the movies *2001: A Space Odyssey* and *Stargate*. A stargate could consist of a series of wormholes connecting many different parts of the universe, much as an old-fashioned telephone switchboard was used to shunt messages from one hotel room to another. However, the engineering problems are akin to those that might be faced by an ancient cuneiform scribe who was suddenly inspired with the idea of a 20th-century word processor. The development of stargates awaits either very exotic materials or ways to create incredibly powerful magnetic fields.[13] Cosmic wormholes are unlikely to be developed, at least by humans, for many generations to come.

Thus, immense distances, expensive fuel requirements, engineering constraints, and political realities are likely to keep us in our part of the solar system for the foreseeable future. We may find some useful shortcuts, but we have not found them yet. We won't run into ET as we wander through another solar system, because we won't be wandering outside ours. Perhaps ET will wander into ours. Yet, extraterrestrials too will have to overcome physical, economic, and cultural constraints and, with 10^{20} stars to explore, pick our own beloved sun. Maybe they will have transportation systems beyond our comprehension, or perhaps as a spacefaring society they will have migrated outward on slow ships until they are spread throughout our galaxy and eventually enter our solar system. If their goal is to communicate with other spacefaring technological civilizations, they will have chosen a very expensive and time-consuming approach.

Robot Probes

Although for us interstellar travel remains in the future, space probes are a reality. Our neighbors may not know this yet, but there are already such probes cruising our part of the galaxy. Two *Pioneer* probes carry plaques bearing engravings of a man and a woman and astronomical reference points indicating the general location and time of origin. Two *Voyager* probes went forth with recordings of terrestrial sights and sounds, along with styluses and playback instructions. The information chosen for export has been criticized because, among other things, the plaque shows woman secondary to man and because the contents of the disk are thought to be ethnocentric or nonrepresentative.[14] Nonetheless, a being that discovered one would find definite proof of intelligent life elsewhere, learn something about human culture, and find where to look for additional details.

A strong case has been advanced for the use of space probes that travel to distant galaxies to propagate information about the sponsor or to seek information about inhabited planets.[15] Such probes could conduct remote observation from orbit or from close observation following a landing. From orbit, probes can identify constructed objects on planetary surfaces,[16] and, as in the case of our own *Viking* probes, landers can analyze surface samples for signs of life.

Consequently, in addition to listening for radio signals, we might look for probes of extraterrestrial origin. Careful estimates suggest that an alien probe in Earth, lunar, solar, or other likely orbits could be located by radar, and that the minimum detectable size would be on the order of 1 to 10 meters.[17] Such probes could be already present in our own solar system; orbital space is *at least* 99.999 percent unexplored for 1- to 10-meter objects. Rapid advances in computer capacity, robotics, nanotechnology, and space exploration should make it possible within 200 years for us to launch our own fleet of space probes, each smaller, cheaper, and lighter than a basketball or a bottle of champagne.[18]

Recently Frank Tipler suggested that whereas relatively small, lightweight, and inexpensive spaceships could not carry humans, they could carry emulations of humans, encoded in computer memory.[19] Tipler believes that we are rapidly approaching the point where the contents of a person's entire nervous system can be encoded in computer memory in such a way as to create a complete and perfect emulation of that person. Sending virtual people offers a number of advantages. We could use a relatively small spacecraft rather than a giant ark to send a population equivalent to that of a small town to the other side of the universe. Radiation shielding and other life-support problems are minimized, and we do not have to worry about the physical effects of acceleration or deceleration on the human body. Finally, the time expended on the mission would be inconsequential to the "participants," since we can compensate for immense mission durations by varying the running speed of the computer or keeping the emulated people engrossed in an interesting virtual world. Would this qualify as a manned mission or as an unmanned mission, and, if it is a manned mission, will it be undertaken by "real" people? There are more than a few philosophical issues here!

Microwave Observation

Much of the information we have about the world around us arrives in the form of electromagnetic energy. Along the spectrum ranging from gamma rays (10^{-14} meters) through electricity (10^{6} meters) is a segment in the vicinity of 10^{-6} meters that produces visible light. This is a narrow segment ranging from about 380 to 760 nanometers (a nanometer is one-billionth of a meter).[20] We are oblivious to waves that fall outside this narrow range except insofar as we can convert them into energy that activates our senses. For example, X rays, which fall below the range of vision, are revealed by their action on photographic plates. Waves that are too long for our eyes are converted into sound and pictures by radio and television. We are continually bombarded by waves within the radio frequencies (10^{-2} for radar through 10^{4} for AM radio) and, as you read this, passing through you are all sorts of radio and television transmissions. Some waves within these frequencies are created by natural processes (such as electrons orbiting in the magnetic field of our galaxy), whereas others are directed by human intelligence.

Although we can travel at but a fraction of the speed of light, radio waves do travel at the speed of light; for most applications radio transmission is instantaneous. Radio makes it possible to communicate at extremely long distances and with many recipients at once. Many other technological civilizations, like our own, may rely on radio communication. Indeed, if in the course of technological evolution radio appears before interstellar travel, then there could be many cultures capable of using radio for each culture that is capable of interstellar travel. By using radio telescopes to conduct microwave searches we might be able to eavesdrop or, more easily, detect a deliberate broadcast intended to attract the attention of beings like us.

During the early decades of the 20th century, when there was some reason to believe that intelligent life might exist on Mars, Marconi and other radio pioneers attempted to intercept transmissions from other planets within our solar system. These early investigators heard nothing unusual, or they intercepted emissions that may have seemed remarkable at the time but cannot be attributed to intelligent extraterrestrials.[21] The first modern search involving a radio telescope and directed at intercepting microwave emissions originating outside our solar system was initiated by Frank Drake in 1960. The 85-foot Tatel telescope that Drake used at the National Radio Astronomy Observatory was capable of detecting, from 10 light-years away, radio signals comparable in strength to Earth's strongest radar transmissions.

Many other searches were initiated during the following decades, and by 1984 Jill Tarter was able to catalog over 40 searches that ranged in duration from just a few hours to hundreds (now thousands) of hours.[22] The number of searches has grown steadily, and as of 1995, major searches were under way in Argentina, Australia, France, India, Italy, Russia, and the United States. Among the more prominent contemporary U.S. searches are META (Megachannel Extraterrestrial Assay), sponsored by the Planetary Society and conducted by Harvard University using a dedicated Harvard–Smithsonian antenna; OSURO, initiated by the Ohio State University Radio Observatory in 1971 and continuing now; PHOENIX, sponsored by the SETI Institute of Palo Alto, California, generously funded by private sources and involving personnel, strategies, and equipment initially associated with NASA's SETI efforts; and SERENDIP (Search for Extraterrestrial Emission from Nearby Developed Intelligent Populations), conducted under the auspices of the University of California at Berkeley. SERENDIP is a parasitic, or piggyback, search that is conducted in the course of other radio astronomy observations. These searches differ in their funding sources, strategies, and technical specifications, but all are based on the conviction that if extraterrestrial civilizations exist, we should be able to detect their electromagnetic activity by means of radio telescope.[23]

Unfortunately, conducting a microwave search is not simply a matter of setting your Radio Shack shortwave on scan and hoping for the best. Because of the vast distances that incoming messages must travel, they will be extremely weak if detectable at all. It is also necessary to conduct a directional search, one that involves aiming a large and sensitive antenna at promising stars. Even as we can improve the reception of a television station by rotating and adjusting the TV set's rabbit ears, we can improve the reception of a very distant radio signal by carefully aiming a microwave dish. A huge microwave dish can detect signals that are perhaps two million times weaker than those that can be detected by rabbit ears. A directional as opposed to an omnidirectional antenna (such as a whip antenna) also makes it possible to triangulate on the transmission's point of origin, although over interstellar distances this will remain a rough approximation. The downside of the directional search is that it takes tremendous patience and time to do the job. Despite the tremendous size of some radio telescope antennas, the amount of sky that can be covered at any one time is about one-tenth the size of the moon, or about the same amount of sky that you would see if you viewed it through a drinking straw.[24]

Defining the Search Space

Embarking on an interstellar journey to search for technologically advanced life-forms is akin to wandering through the most remote parts of the Sahara and hoping to find your next-door neighbor. Even optimists believe that the number of civilizations will be few relative to the number of stars. Even with a thorough search they will be difficult to find, because on average, technologically advanced civilizations are likely to be a few hundred to a few thousand light-years apart.[25]

Given billions of galaxies, each containing billions of stars, and given millions of radio frequencies, where do we look? What types of signals should we seek? Defining the search space, that is, identifying stars and frequencies in such a way as to increase the probability of success, is an exercise in human decision-making.

Rational, or "econologic," decision-making models assume that people can identify all the alternatives, think through the implications of each alternative, then make a fully informed decision. These are prescriptive models of decision-making, ideals rather than realities. In real life we simply lack the time and information to follow this procedure. Instead, we have to limit the amount of information that we consider and take shortcuts to find workable solutions. Rather than beat the bushes for a full range of alternatives, we focus our attention on areas we think show the greatest promise. To do this, we follow rules of thumb known as heuristics.[26] The possibilities we overlook and the shortcuts we take occasionally lead to failure or force us to get by with make-do solutions. At the same time these efficiencies help us survive daily life and to attempt tasks that would otherwise be unmanageable. In SETI, heuristics are used to identify the locations in space, the radio frequencies, and the signal types that are the most promising for yielding evidence of extraterrestrial intelligence.

Any one of countless stars could host a planet with a technologically advanced civilization. However, the stars outside our galaxy are so far away that it is unlikely that even a deliberate interstellar beacon would be so powerful as to reach us in intelligible form. In addition, the distances are so great that hundreds of thousands (or, more likely, millions) of years would have passed since the signal was transmitted. (Remember that the next galaxy is over 75,000 light years away.) Because of such distances, real two-way communication seems impossible. So the first rule of thumb is to focus on stars within our own galaxy. As noted in chapter 1, not all stars are equally promising hosts for life; the most promising are those that are similar to our sun. SETI searchers do not wish to dismiss all other possibilities, so some searches concentrate on stars like ours but also scan less promising neighbors.

There are literally millions of radio frequencies that might carry messages from outer space. If we assume that their experiences with radio are similar to our own, and that they are trying to reach civilizations such as our own, then certain frequencies can be discarded. Two obvious heuristics are to ignore frequencies that don't carry signals all that far, and to forget about especially noisy (static-filled) portions of the radio band.

The next heuristic is based on the controversial assumption that their scientists will think like our scientists. This involves the idea of a "cosmic water hole," a radio frequency where intelligent civilizations might gather to communicate with one another. The cosmic water hole is a frequency band between 1,400 and 1,800 megahertz (that is, a wavelength of 18–21 centimeters). This, the least noisy part of the electromagnetic spectrum, includes the "song of hydrogen" and a few other special frequencies. Hydrogen, the most abundant element in the universe, radiates at the wavelength of 21 centimeters.[27] Any civilization that has begun exploring the universe must have detected this universal radiation and realize that other civilizations will recognize it too. This frequency range thus provides an obvious and convenient place for societies to gather.

Anyone who has ever tuned a shortwave knows that radio bands are replete with all kinds of yowls, whistles, screeches, and static as well as voice communications of almost every imaginable form. Embedded within these sounds may be those of another civilization. As we focus our microwave dishes on the heavens, naturally occurring signals, the cosmic equivalent of static, will predominate. However, naturally generated radio waves tend to be wide band in the sense that they cut across several separable frequencies. Consciously controlled or deliberate signals are usually quite narrow, occupying but a small slice of a radio band. The task is one of separating narrow signals, which rarely occur in nature, from wide-band signals, which are abundant in nature.

After identifying narrow-band signals, searchers must eliminate those that are not of terrestrial origin. To some extent, this can be achieved by the use of highly directional microwave equipment. There remain many distractions (such as transmissions from passing aircraft), but in most cases these red herrings are easy to identify because of their known characteristics. Then there are the narrow-band transmissions of extraterrestrial origin that result from natural physical processes not under intelligent control. Searchers hope that civilizations advanced in the ways of astronomy will choose a signaling method that does not mimic nature.

Hits and Misses

For search purposes, radio interceptions can be put into one of four categories. These categories represent combinations of two dimensions. One dimension dichotomizes the interceptions themselves: those emanating from intelligent extraterrestrial entities and those emanating from all other sources, both naturally occurring and human made. The second dimension represents our interpretation of these interceptions, that is, whether we attribute them to extraterrestrial intelligence (positives) or believe that they are something else (negatives). Let us begin with the least interesting case.

True negatives are normal everyday interceptions that are unrelated to extraterrestrial intelligence and whose irrelevance (for purposes of SETI) is recognized. All but the tiniest fraction of microwave interceptions fall into this category. In most cases, commercial radio broadcasts, military and ham communications, emanations from industrial microwave ovens, and the like are easily identified and appropriately eliminated from further consideration.

False negatives are missed opportunities. They are interceptions that are in fact of intelligent extraterrestrial origin but are dismissed as something else: *Star Trek*, the kid next door experimenting with a radio kit, sunspots, and so forth. These interceptions are ignored when they shouldn't be, but unless we are given another chance, we'll never discover our mistake. There is no way to estimate how many communications from outer space (if any) have been ignored or dismissed.

The radio bands are crowded with a cacophony of whistles, tweets, rumbles, and unintelligible speech. Try experimenting with a shortwave radio and see how well you can identify various signals. Even people who are well versed in such matters intercept many mysterious transmissions, although the use of radio direction-finding equipment and other means suggests that they originate on Earth. For example, a shortwave listener in the early 1980s could find many unexplained signals, including a 4-note electronic tune (3,280 kHz); a plaintive, 21-note flutelike melody (12,700 kHz), an overpowering signal variously described as a woodpecker or buzz saw (3,261–17,540 kHz), and transmissions that some commentators thought were part of secret "death ray" experiments.[28]

False positives are false alarms—interceptions that someone thinks signify extraterrestrial intelligence but that are something else. These cause a brief stir of excitement, followed by disappointment when their nature becomes clear. False alarms can come about in a number of ways. A deliberate hoax is one cause. For example, one way an alien might signify its presence is to reflect back to Earth one of Earth's own signals. The idea is that if we receive one of our own signals sometime after it is broadcast, we will realize that the echo is under intelligent control. Seemingly, just this type of signal was received in England in 1953. It was a test pattern from KLEE, a television station in Texas. Not only was this test pattern received at an incredible distance, it seemed to have been lost in time, because several months earlier

KLEE had been assigned new call letters. Where had the signal been and how did it arrive in England from Texas? Could KLEE's signal have been captured and rebroadcast to Earth? At the time, a group of British businessmen were urging investors to back "special" television sets said to be capable of receiving programs from great distances. To demonstrate this, potential investors were shown sets that displayed call signs or test patterns from around the world, including station KLEE. In fact, the demonstration was fraudulent, and the test patterns were projected onto the television screen using a slide projector.[29]

A second kind of false alarm is the misinterpretation of a naturally occurring extraterrestrial phenomenon as being under intelligent control. For example, in the mid-1960s, radio astronomers identified signals from the far reaches of the galaxy that were notable for their incredible cyclic timing. As reported in the *New York Times* and other news outlets, some scientists initially believed that these signals represented extraterrestrial navigation beacons. Almost immediately scientists found a more prosaic answer: the emissions came from pulsars, highly interesting but entirely natural phenomena (see chapter 9).

A third type of false alarm is based on the misidentification of artificial phenomena. This type of misinterpretation may be a special risk for those of us who are hoping for evidence of extraterrestrial intelligence. The wish combines with ambiguous evidence to produce a false positive. The same psychological process of projection that allows us to see faces in clouds or to make up stories about pictures or inkblots causes us to misinterpret what we see or hear.

A good example of wishful thinking comes from Philip Hough and Jenny Randles, who describe a woman tuning a shortwave radio in the early 1960s.[30] Although reception was poor and parts of the transmission were garbled, both the woman and the people who listened to her tape recordings of the broadcast agreed that it sounded like "This is a test transmission for circuit-adjustment purposes from a radio station of [garbled]... This station is located in outer space." After reading about this in a journal devoted to unusual phenomena, one resourceful individual wrote that he had filled in parts of the message with a beat frequency oscillator (an electronic augmentation) and had clearly heard "This is a test transmission for circuit-adjustment purposes from a radio station of the Hellenic Communications Organization. This station is located in Athens, Greece." The original listener's imagination filled in an indistinct message, and her interpretation created expectations in subsequent listeners, whose own imaginations confirmed her discovery. When the transmission was augmented electronically, the real content was unmistakable.

The *true positives* are in the "Bingo!" category: signals of intelligent extraterrestrial origin and correctly interpreted as such. This category is empty: whatever their private suspicions, scientists are not willing to admit that any interceptions belong here. In radio telescope searches conducted over the years, there have been many suspicious signals, but they either turn out to be false alarms or cannot be verified. During the first day of the first search, in about 1960, Frank Drake was rewarded almost immediately with a highly provocative signal, but careful attempts at verification revealed that it came from a passing jet plane.[31] The search at Ohio State University yielded a very powerful signal that had all the requisite characteristics, yet it was never demonstrated to be of extraterrestrial origin. This exceptionally strong signal is called the WOW signal because of the observer's initial reaction. It has never been found again.[32]

It is interesting that with over 30 years of searching under our belts, and with the rate of the search accelerating as new technology comes on-line, almost all the false alarms discussed in print occurred two or three decades ago. False alarms still occur, but searchers seem reluctant to mention them. Publicizing false alarms may trigger rumors, suggest that the searchers do not know what they are doing, and will make it difficult to convince the public that contact has actually occurred once a real signal has been intercepted. In addition, the disclosure of promising but unconfirmed signals attracts reporters who push for premature confirmation and, after failing to get it, may harbor dark suspicions of a cover-up.

The most important reason that scientists are reluctant to ascribe unusual signals to intelligent extraterrestrials is scientific. A scientific observation is a repeatable observation. Thus, promising candidates must be verified by independent observers before success is proclaimed. Consequently, we hope that either the extraterrestrials keep transmitting so that following the initial detection they can be found again, or that several listeners happen to tune in simultaneously. The problem with signals like the WOW signal is that they are heard only once. Science is a slow and cumulative process that involves the careful assemblage and verification of evidence. Even the most tantalizing signals must be considered false alarms unless there is proof to the contrary.

Limitations of a Radio Search

Whereas heuristics may steer us in promising directions, they also move us away from other possibilities. Consequently, the same rules of thumb that make the search manageable also limit its scope and may keep us from locating our Galactic neighbors. Certain stars, certain frequencies, and certain types of messages are being sought; other stars, other frequencies, and other types of messages might bring us more exciting results.

A radio telescope search rests on many assumptions. Foremost among these is the assumption that advanced technological civilizations communicate by means of radio. This may not be true, at least in the sense that their radio activity may not satisfy our requirements for a successful microwave search. Although for operational purposes we define as advanced technological civilizations those that use radio, there may be highly advanced civilizations where radio as we know it does not play a major role: an undersea civilization for example. Or radio may be discovered fairly early in the course of technological development, but if a civilization moves on to other means of communication, then it may no longer be detectable by radio.

Yet another worry is that extraterrestrials may use information-compression technology. This involves compressing a lengthy message into a brief transmission, or reducing bandwidth (saving air space) by broadcasting only parts of a message to a receiver that is able to electronically reconstitute the original message. Information compression reduces the problem of crowded airwaves, a problem that we might expect to confront in extraterrestrial civilizations. Messages that are brief or that take relatively little frequency space will of course make them all the more difficult to find.

Another limiting assumption is that extraterrestrial civilizations will transmit on a particular wavelength, specifically in the vicinity of the cosmic water hole. Whereas the water-hole hypothesis provides a place to start, it is by no means assured that the reasoning that led us to the water hole would lead them to the same place. Indeed, papers contributed to recent volumes edited by Seth Shostak offer a plethora of other frequencies as logical places to search.[33] Yet, magic frequencies may soon become a non-issue as search technology improves by leaps and bounds and the search encompasses more and more frequencies. By the mid-1990s, receivers scanning over 8 million frequency bands were in operation, and receivers with billions of channels loomed on the horizon.[34]

We can torture ourselves with an endless list of things that might go wrong in a microwave search. They transmit in one direction, we listen in another. Our receiver's frequency range ends just short of their transmitter's frequency range. Their transmitter is just a smidge too weak or our receiver is just a mite too insensitive. If only our microwave dish had been just a silly meter larger! They stop transmitting a second before we turn on our receiver. We turn off our receiver a second before they start transmitting. And so on, and so forth.

Their motives will affect our chances of success. Intentional beacons are likely to be of exceptional power and hence capable of traveling interstellar or even intergalactic distances. Additionally, transmissions would be structured in such a way as to attract attention and to be decipherable by creatures of many intelligence levels. Detection through "eavesdropping" is not particularly promising. Since messages not intended for an interstellar audience are likely to be weak, we probably will not be able to detect them. Furthermore, if the messages are not intended for "outsiders," the broadcasters will have taken no special steps to ease our attempts at decryption. Of course, failure is guaranteed if ET seeks to maintain privacy. Even today's terrestrial technology encompasses many methods that could make a high-powered message impossible to detect. We already have "screech transmissions," consisting of tape-recorded messages speeded up or scrambled in such a way that they sound like radio interference, as well as "frequency hopping" transmitters and receivers, where two synchronized units broadcast and receive on one of more than a score of frequencies that change every few seconds.[35]

Even with a strong commitment and steadily improving technology we may have to search for centuries. We might expect a succession of searches, each fueled by new technologies. There are at least three ways to improve our chances of contact.[36] The first is to use several radio telescopes at once. Each of these scans the same frequencies and locations at the same time. Even as several sailors might be better equipped than an individual lookout to spot a whale or an enemy ship, a group of radio telescopes is better positioned than an individual telescope to detect something interesting in the heavens. Furthermore, the "oversampling" made possible by an array of receivers tends to minimize the contribution of noise to the overall signal. During the 1970s, a NASA planning exercise resulted in Project Cyclops, which was to involve a field of a thousand 100-meter radio telescopes. The awe-inspiring artist's rendition of this incredible array of microwave radio dishes diverted attention from the project's sensible strategy: to begin with only one radio telescope and then add more telescopes as needed.[37]

Our current SETI involves searching from Earth. Unfortunately, atmospheric conditions and a high volume of radio traffic make it difficult. Although certain radio frequencies are legally protected for scientific purposes, terrestrial transmitters, seeking to operate in an increasingly crowded radio spectrum, frequently impinge on channels reserved for passive listening.[38] Thus we might want to search from a high-orbit satellite or from inside a crater on the back side of the moon, locations that would protect radio telescopes from the kinds of interference that plague a search on Earth. Later still we might search from the far reaches of our solar system, with very large solar sails serving as directional antennas.[39]

Yet another procedure is locating the radio telescope in deep space, at a specified distance from another star. This star, interposed between the transmitter and the receiver, would serve as a "gravity lens," which in effect magnifies a distant signal. With this procedure we might be able to detect civilizations in other galaxies. Meanwhile, we can expect the continual upgrading of our antennas, receivers, and data-analysis techniques. Over the past few decades we have become able to simultaneously monitor tens, hundreds, thousands, millions, and soon billions of channels. There have been similar advances in the computers and computer programs used for separating promising interceptions from all other kinds.

Proponents of alternatives that are not based on microwave observation believe their strategies overcome some of the weaknesses of radio searches. Robert Zubrin, for one, suggests that we need not assume that extraterrestrial cultures use radios at particular frequencies if we seek the spectral signatures of their interstellar spacecraft.[40] He argues that we do not need to understand alien psychology to detect interstellar travel, which is strictly governed by the laws of physics. One of the many justifications for developing interstellar craft, notes Robert Forward, is their ability to conduct searches for extraterrestrial intelligence that are in some ways more thorough than can be conducted by radio alone.[41] After all, the discovery of any extraterrestrial life-forms will be monumental, and many life-forms will not have radio. Of course, alternative search strategies are based on their own assumptions (such as those about the "typical" star ship) and have their own shortcomings (for example, it would take considerable time for interstellar explorers to redirect their efforts from one solar system to another). Certainly we should be open to many strategies. Yet the compelling advantage of a radio search is that it is possible to rapidly search immense volumes of space using technology that is on-line right now.

CONTACT AND COMMUNICATION

Current microwave search strategies involve locating a carrier wave, that is, a "pure" radio wave, under intelligent extraterrestrial control. Apart from announcing its presence, a carrier wave by itself gives little information: it is comparable to a dial tone. The actual message (if any) is superimposed on it. The search equipment that makes it possible to identify the carrier wave will not reveal the superimposed information; after the carrier wave is found, the searchers will have to look again with different equipment if they want to decipher the transmission or exchange information.

We hope, therefore, to find some basis for understanding the superimposed messages. Biological differences in the form of anatomy and physiology and especially neuroanatomy (or its equivalent), and cultural differences could make it difficult for an exchange of information to take place.[42]

Biological Factors

Different biological organisms have different morphologies and structures. Even though cross-species comparisons of certain terrestrial structures reveal a degree of functional equivalence, there remain differences that may have important implications for alien–human communication. For example, although all living organisms process information, some have better developed or more complex nervous systems than others do. Thus it may be impossible for the less developed organism to grasp abstractions and relationships that the more developed (or at least different) organism instantly comprehends.

Organisms that have different hard-wiring may find it difficult to communicate simply because they do not share the same experiences. Douglas Raybeck illustrates this with an analogy: a conversation between a color-blind and a color-sighted person. The person who can see colors has experiences and uses metaphors that are beyond the realm of the person who is color blind.[43] In a similar vein, crabs, fish, insects, and many other creatures have eyes that seem primitive by human standards. Their eyes lack the power of resolution required to distinguish between letters of the alphabet. Could we be similarly disadvantaged in our attempts to understand messages from more advanced species?

Communication with extraterrestrials may run afoul of differences in time, scale, and complexity. For some creatures, time passes faster than it does for others. Compare, for example, the rates of movement and the life spans of the hummingbird and the elephant. A substantial period of time for the hummingbird may be a brief interlude for the elephant. If the hummingbird and the elephant were able to converse, the hummingbird's message might be too brief to be understood by the elephant. Similarly, the elephant's message could be so long that it would fail to gain the attention of the hummingbird. The same burst of sound might sound like a single note to one organism, a digital data stream to another, and disconnected discrete sounds to a creature that runs on yet another internal clock.

Under some conditions, difficulties can arise if the communicants are of radically different sizes. Messages formulated by one party may be incomprehensible to the other party simply because, from the second party's perspective, the messages are too big or too small. If you were a tiny insect crawling on a photograph on the front page of the newspaper, the best you could hope to see at any point in time would be a few isolated dots. The overall picture would be impossible to grasp. If you were very small and prepared a picture for creatures your size, the individual dots could merge in the eyes of a very large organism. In the one instance it is impossible to see the forest for the trees; in the other instance it is impossible to see the trees for the forest.

The biggest potential problem is posed by differences in the ability to handle informational complexity. On Earth, we humans seem convinced that we have intellectual advantages over all other species, but we are less confident that we will remain "number one" when our galactic neighbors go on the air. Concerns here include the rate of transmission (and you think terrestrial foreigners speak fast!) and the number of components transmitted simultaneously. It would be easy to overload our neural circuits. Human communication is multicomponent, as it has both verbal and nonverbal components, but with few exceptions (such as Chinese) the words stand alone. This may not be the case for alien languages, which could involve multiple harmonics or tones to convey even simple ideas. As John Baird has pointed out, length of message is another important consideration. Some messages may simply be too long for us to comprehend; it is doubtful, for example, that we could break up a 300- or 400-word "sentence" into manageable and meaningful chunks.[44]

Message complexity poses certain dilemmas for the beckoning civilization. On the one hand, a simple message can be identified, but it might be mistaken for natural phenomena and it conveys little information. On the other hand, the complex message contains tremendous amounts of information but may be too difficult for the recipient to understand. Thus, Ian Ridpath has suggested that any civilization attempting to contact creatures like us would do well to send three levels of messages simultaneously.[45] At the first or most superficial level we might find an announcement of their presence in the galaxy and an invitation to explore more detailed transmissions. At the second level we might learn about the broadcasters, including an explanation of their civilization, culture, and technology. This would provide sort of a *National Geographic* overview of what they are like. At the third and highest level we might discover highly detailed descriptions of the life and history of their culture: an encyclopedia suitable for scholars. Broadcasting on all three levels simultaneously minimizes the immense time lags involved in a question-and-answer exchange.

Social and Cultural Factors

Members of terrestrial and extraterrestrial cultures live in one universe that is governed by the same laws; an atom of oxygen is the same everywhere. However, the assumptions that terrestrials and extraterrestrials make about the universe, the models that they devise to portray it, and the ways in which they think and talk about it may be very different.[46]

It is useful, at this point, to draw a distinction between two kinds of meaning. *Denotative* meaning refers to actual objects and events. It is, in effect, bare-bones objective meaning. If we say that Klong threw the gauntlet on the table, the denotative meaning is that Klong took off his glove and threw it on the table—nothing more.

Connotative meaning is additional or even substituted meaning that results from mental association and with instruction from other people. Connotative meaning encompasses the images and feelings that the message prompts. For many of us in western society, "Klong threw the gauntlet on the table" does not mean that Klong threw a glove, but that Klong issued a challenge. Indeed, there may never have been a glove or a table or even (in Klong's case) a hand. The power of connotative meaning is revealed in our reactions to "ET" and "alien." Although we can use these words interchangeably to designate an intelligent extraterrestrial, in the absence of additional detail they are likely to trigger very different images and emotional responses.

Connotative meanings are culturally conditioned and therefore vary from culture to culture. An important connotation in one culture may be absent or radically different in another culture. Raybeck notes that even simple pictures may miscommunicate.[47] In contemporary western societies a heart with an arrow through it is a sign of love, but it may have less pleasant connotations to a viewer who sees only physical damage. The annals of anthropology are replete with cases in which differing connotations either preclude communication or lead to unintended consequences.

Typically, languages draw on local referents, that is, objects and relationships that are unknown to another culture. Indeed, because of these referents people from different regions within the same culture may have difficulty understanding one another. Illustrative here is the lingo of Boontling developed in Booneville, California, between the 1880s and the 1930s.[48] Located between the northern inland community of Ukiah and the Mendocino coast town of Fort Bragg, Booneville was somewhat inaccessible. Its Boontling lingo was internally consistent, with words that were agreed upon by all who helped develop it.

Because so many of the words' meanings came from purely local characters or episodes, Boontling was difficult for outsiders to understand. Indeed, one of the functions of Boontling was to protect communications in the presence of outsiders. Boontlers could discuss a business deal in front of the San Franciscan who proposed it without this person understanding the full depth of their analysis, or they could engage in bawdy talk ("nonch harpin") in front of someone whose only reaction was curiosity. Of course, people from Booneville also understood standard English, but if Boontling had been their only language it would have been difficult to establish satisfactory communication with outsiders.

FACILITATING COMMUNICATION

The problems of communication are formidable. First of all, we must hope that ET's neural wiring is such that it offers hope for our understanding each other. Second, we have to hope that our experiences overlap enough to provide a basis for shared meaning, not only in the physical sciences but in other areas too. For such reasons we can expect anthropological linguists and cognitive psychologists to be heavily involved in solving communications problems.

Deciphering a message from long-dead terrestrial cultures is in some ways analogous to deciphering a message from outer space. In each case we try to understand beings that are very distant from us. Ben Finney and Jerry Bentley point out that SETI scientists are heartened by the transmission of Greek science to medieval western Europe, a process that involved ancient Greek philosophers and scientists, their Arab and Byzantine successors, Jewish translators, and western European scholars. Despite this enormous success, note Bentley and Finney, we have to recognize that the many people and generations involved were cultural cousins who could easily learn one another's languages. Let us look at a case where there was a greater gulf between the people who composed the messages and those who deciphered them: the ancient Mayan culture and the 20th-century scholars who tried to translate its writings. These writings, which are reminiscent of Egyptian hieroglyphics, are found in parts of Mexico, Guatemala, Belize, and Honduras, and date from about 250 A.D. to about 900 A.D.[49]

For years, attempts to understand Mayan writings were thwarted by the belief that the inscriptions were ideographs that expressed ideas directly, that is, without letters, words, or sentences. In fact, note Finney and Bentley, no matter how picturelike they appear, the glyphs express phonetic syllables and morphemes (the smallest units of meaning). As in the case of every other decipherment of ancient writing, the breaking of the Mayan code required some understanding of the writer's language. Since we will have no understanding of ET's language, how can we hope to understand ET's message?

There are many, sometimes wishful proposals for facilitating communication between humans and extraterrestrials. These include mental telepathy, pictures, developing a universal language, and computer-aided translation. Some of these are proposed in the context of how we might communicate with ET, but the same procedures would facilitate ET's communication with us.

Mental Telepathy

Through clairvoyance, psychics hope to form mental images of life in distant galaxies, or through telepathy communicate with beings on far-off worlds. Certainly mental telepathy would be an attractive medium. It does not require a fancy spaceship and huge amounts of energy; it does not even require radio technology. Distance doesn't matter, and, in some discussions, telepathy would be instantaneous. Such communication cuts across language barriers: rather than beaming unintelligible words, one can beam direct meanings. Over the years, many individuals and organizations have claimed to be in telepathic contact with extraterrestrial civilizations.[50]

Despite its seeming ease and efficiency there is good reason to reject telepathy as an avenue for communication. When we rely on something other than our known senses, our experiences and conclusions cannot be verified by others. Our inner experiences may be quite compelling to ourselves, but when we hear others report mental contact with aliens we cannot eliminate the possibility of error or fraud. We might be persuaded that Earth was in telepathic contact with denizens of another planet in two ways. One would be if the aliens went "public" by communicating simultaneously with all of us, not just a select few. A comparison might be drawn here between a private telephone call, about which each party can make claims the other can deny, versus a public broadcast, where everyone is privy to the same information. The second way would be for the alien civilization to provide us with stunning insights that we could grasp and verify: an elusive mathematical proof, the solution to a scientific problem, the blueprint for a "miraculous" invention (see chapter 3). There is no compelling evidence that either of these events have ever occurred.

Pictures

Pictures are potentially useful, for they give us the opportunity to express ideas directly. The hope is that the direct representation will be sufficiently realistic to be immediately understandable. Adrian Berry has recently described an electronic mail system known as Le-Mail (for Language Education, Monitoring, and Instructional Learning), which consists exclusively of icons. This allows people worldwide to communicate despite tremendous differences in languages. Thus far, Le-Mail is used for making travel arrangements. Of course, the icons are easily understood by humans: adult, child, and infant figures; single and double beds; calendars; happy faces and sad faces.[51]

As already noted, probes bearing pictures have left our solar system, and Earth has transmitted a picture in the direction of the Hercules star cluster in M13. The off–on signals are intended to be arranged in a series of pixels that form the picture, perhaps on a cathode ray tube, perhaps on a printed matrix of dots, perhaps by some technology unknown to us. The trick in decoding such pictures is to identify the number of columns and rows; once this is done, it should be quite easy. (Even if the number of pixels is fairly large, modern computers can rapidly run through all possible combinations until the right number of columns and rows is found, but to me an equal number of columns and rows, producing a square picture, might not be too bad a place to start.) Pictures could tell us where a transmission originated as well as show us local geography, life-forms, and artifacts.

Douglas Vakoch proposes several procedures that we, or they, might use to facilitate communication by pictures. First, keep formatting problems to a minimum. A representation that is self-evident (in that it requires no formatting at all) is best. Second, representations should bear as close a relationship as possible to their physical referent. For example, a picture of Abraham Lincoln would be more informative than a picture spelling out Lincoln's name. We should use icons, which have an intrinsic relationship to what we are trying to express, rather than symbols, which are arbitrary or based on conventions rather than on an intrinsic link. Third, even though we and they live in the same universe, our representations of the universe may be quite different, so that the model of an atom (for example) that makes sense to us may not make sense to them.

To illustrate this approach, Vakoch shows how electromagnetic radiation can be used to communicate chemical concepts. Each chemical element gives off a particular pattern of frequencies. We can mimic these frequencies by transmitting on multiple frequencies and expressing differences in terms of the frequencies and intensities of our transmissions. Thus, the structure of the signal bears a close physical resemblance to the concept that is being communicated, and the recipients are pointed not toward our models of these phenomena but toward the phenomena themselves.[52]

Universal Languages

There are thousands of languages in use on Earth at the present time. These evolved in geographically dispersed societies, which shows, incidentally, the human proclivity to use speech. Since so many intercultural problems rest on miscommunication, wouldn't it be nice if there were some sort of overarching, or universal language that could be used by all?

There have been several attempts to develop a language that could be easily used by people of many different mother tongues. Probably the best known of these is Esperanto, a language that draws from English, Spanish, German, and other European tongues. Developed in the 1800s, Esperanto was intended to ease discourse and smooth the way of the traveler. It was simple in form and avoids many of the idiosyncratic complexities that encumber most natural European languages.

Here we are, over a hundred years later, and Esperanto seems to be used mostly in conversations among members of an Esperanto club. Its failure to take hold reflects the limitations of the language itself (the annoying irregularities and exceptions that make a nonuniversal language so difficult to learn also give it rich meaning) and a general sense that an artificial language is simply not worth learning unless everyone else is willing to do so. Introducing an artificial language is like trying to improve the typewriter keyboard or replace current television sets with incompatible but superior units: the past investment makes one think twice.

Perhaps an interest in conversation among the stars will give universal languages a new lease on life. This is the hope of the Dutch mathematician Hans Freudenthal, who proposes *lingua cosmica*, or Lincos, for this purpose. Lincos would be broadcast from Earth in an attempt to find other civilizations.[53] It is an abstract, formalized language intended to be understood by beings that are not familiar with any terrestrial languages—a problem, Freudenthal notes, that is confronted by babies and infants! In the case of infants we can rely on showing and doing, but this luxury will not be available when it is time to communicate with extraterrestrials.

Lincos involves unmodulated wavelengths like those associated with "cw," or continuous-wave (i.e., radiotelegraph code), communications. Meaning is imposed by varying the frequency and duration of the signal. These variations produce the basic sound units (phonemes) of the language. Pauses, because they are "self-evident," substitute for punctuation. Lincos is a "spoken" language; although there is a graphic or written version, it corresponds imperfectly with the spoken version and is primarily for internal (terrestrial} purposes.

Because it will be necessary to teach Lincos, we must proceed in formal, abstract ways that communicate facts already known to the receiver. Thus, in keeping with a long tradition, Freudenthal proposes beginning with mathematics. In this way it will be possible to introduce various relationships (+, −, <, >, =). Many specific instances are given so that the receiver can grasp the general principle. Mathematics is only the beginning; the idea is to slowly expand the vocabulary, being sure to reiterate old examples as we introduce the new.

A highly abstract language, Lincos rests on as few assumptions as possible. C. L. DeVito and R. T. Oehrle argue that we can make certain simplifying assumptions because contact will be made via radio.[54] Given that the extraterrestrials will possess radio technology, it is reasonable to assume that they will have detailed knowledge of the physical universe and will have investigated the fundamental properties of matter and energy. Members of all societies that develop radio telescopes "can count, understand chemical elements, are familiar with the melting and boiling behavior of pure substances, and understand the properties of the gaseous state"[55] and should be able to learn a language based on fundamental science. Although we must begin with simple goals, it will be possible to exchange complex information before the language is complete.

In stage 1 of DeVito and Oehrle's procedure we assume only that ETs can count and understand natural numbers; we begin by repetitively presenting notation for shared concepts. In stage 2, we introduce non-trivial mathematics. In stage 3, "matter," we discuss the physical universe, beginning with the set of chemical elements. We can provide a number of clues so they know that the message deals with elements. In stage 4, a discussion of heat turns into a discussion of energy, and in stage 5 we move to more advanced mathematics, including solid geometry.

Computer-Aided Translation

Some of the people who allege they have had contact with aliens report that an electrical device was used to permit communication. Such contraptions often consist of a small box mounted on the front of the alien space suit, garment, or skin. This computer (or whatever) seems capable of translating any language on Earth to a language accessible to the aliens, and vice versa. Whereas we may question the accuracy of such reports, computer-aided translation itself is not far-fetched.

Certainly, in our own time, we see remarkable advances in such pivotal areas as computer voice and pattern recognition and, of course, in computer power. Already there are electronic devices capable of dictionary substitutions and of translating simple phrases. These devices are available to globe-trotters for less than a hundred dollars. Although translations into unknown languages are another matter, Raybeck proposes that new types of computer architecture might provide a base for interstellar communication.[56] Essentially, computers would bridge the gap caused by dissimilarly wired nervous systems.

Such a computer would be instructed in a language, beginning with basic rules and a basic vocabulary and then work up to more complex communications encompassing subtle shades of meaning. The computer would be instructed to search the other language for identifiable patterns. In effect, it would serve as a higher order, or superordinate, intelligence, one that comprehended the vocabulary, grammar, and implications of both languages, and it would attempt to frame messages within each culture's, customs, and traditions. Computerized hard-wiring could also accommodate variations in preferred communications modalities, for example by converting sounds into printed words into pictures, and so forth.

When all is said and done, we may be rescued by our own mediocrity, the likelihood that, technologically speaking, they are more advanced than we (chapter 1). Technology involves more than spacecrafts and advanced communications devices; it involves instructional and learning technology. An older and wiser civilization is likely to have taught other species how to communicate and, particularly if its members have done this many times, will be able to use its findings to communicate with us.

We ourselves have come a long way in our own instructional technology, including computerized instruction. This form of instruction can be devised to accommodate differences in initial knowledge and competence. Even toy computers are able to adapt their games to the skill level of the player, and as the player progresses, he or she can advance to more challenging and potentially more rewarding levels. Perhaps such technology has already been applied to interstellar communication.

CONCLUSION

Whereas other advanced societies may have mastered traveling interstellar or even intergalactic distances in acceptably brief periods of time, there is no compelling evidence that they have visited Earth. We ourselves will lack the ability to travel outside our solar system for many generations to come, maybe forever. The most feasible way for us to find signs of intelligent life elsewhere is through microwave observation, that is, intercepting microwave transmissions that are of extraterrestrial and intelligent origin. Such searches have proceeded for about 40 years and have involved the analysis of hundreds of trillions of interceptions. Despite the occasional tantalizing lead, there have been no verified signals from an extraterrestrial civilization.

The same period of time has seen tremendous growth in our technical ability to conduct radio searches. I am more optimistic about the continued expansion of our technical ability than I am about our capacity to sustain our motivation long enough to get results. The U.S. Congress seems incapable of pursuing programs that last beyond the next election, and the search may have to go on for an unnecessarily long time because of fluctuations in public support.

Could North American society maintain a multigeneration search? In times past people were willing to work on projects that would not bear fruit in their lifetime, for example, a cathedral that might take hundreds of years to complete. Contemporary North American society—with its emphasis on personal responsibility, achievement, and individual performance—may not be the ideal incubator for people who are willing to spend their lives on projects with an uncertain time line and outcome. (Asian societies are another matter. There is an emphasis on group responsibility and achievement, and a willingness to set goals many years ahead.) There may always, however, be a few North American philanthropists who are willing to support the search, perhaps even contribute to a permanent endowment, and a few scientists willing to persevere, at least on a time-share basis.

THREE

FALSE ALARMS

Late June 1947 was an exciting time for aviation enthusiasts. A huge military air armada was touring the Northeast. Several commercial aviation records had been set, and the newspapers discussed the amazing array of fighters and bombers coming on-line. On June 25, the B-50 bomber was tested near Seattle. Aloft for 1 hour and 38 minutes, it achieved speeds approaching 400 miles per hour.

Also near Seattle, that same day, Kenneth Arnold, a businessman from Boise, Idaho, identified in the press as a "veteran pilot and fire control engineer," sighted "nine saucer-shaped objects dipping and skimming through the sky" and flying in a "weaving formation" at about 10,000 feet. By clocking the formation's travel time between Mount Rainier and Mount Adams, Arnold, who was piloting his own plane, calculated their speed at about 1,200 miles per hour. Asked to speculate on their nature, he stated, "I don't know what they were—But I know that I saw them." He reportedly clung "stoutly to his story" and bought an enormously expensive camera to photograph the saucers if he encountered them again.[1]

In his efforts to provide an accurate description Arnold

helped coin a term that became a permanent part of our language: "flying saucers." Although later a more neutral and encompassing descriptor was introduced—*UFO*, for unidentified flying object—the term "flying saucer," with its humorous connotations, has been with us ever since.

For centuries there have been reports of unusual lights and objects in the sky, but Arnold's sightings ushered in a new era of claims. According to one hypothesis, such sightings—or at least some portion of them—are evidence of intelligent non-human beings from outer space. If one accepts this hypothesis, then the whole SETI enterprise seems superfluous. Why search the heavens with a radio telescope if ETs are already here? Consequently, it is useful to pause at this juncture to consider the claim that we are regularly visited by beings from outer space. The results of this consideration will help explain why, despite a common fascination with extraterrestrial life, UFOlogists (people who study UFOs) and SETI scientists walk separate paths.

A BRIEF OVERVIEW OF UFO-RELATED PHENOMENA

There are literally hundreds of books on UFOs, and as of October 1996, there were approximately 20,000 references to UFOs on the World Wide Web.[2] The books report, with varying degrees of conviction or cynicism, thousands of sightings and encounters of a closer kind. There are statistical summaries, brief case histories, and extended discussions. Reports include those of eerie lights (in some cases reminiscent of the light cast by a welding arc); spherical, conical, cylindrical, ring-shaped, and top-shaped as well as disk-shaped objects ranging in size from a few feet to miles in diameter; and multiple objects performing intricate drill-like maneuvers. The objects may be anywhere from a few feet to many miles away. There is almost every kind of witness that one can imagine (including intelligent, honest witnesses), and sighting duration ranges from seconds to hours. People have reported seeing UFOs landing and occupants performing repairs or collecting botanical specimens. On occasion the observer is invited to peer into or even enter the craft, and more than one person reports having been offered a bland but healthy extraterrestrial snack.

Since the 1947 sightings, the "UFO myth," notes Curtis Peebles, has waxed, waned, and shifted direction, but the central idea—that we are being visited by intelligent extraterrestrials—has persevered.[3] Over the first half decade or so, an avalanche of reported sightings, governmental agency reactions, news media accounts, supposed mysterious happenings, popular books, and science fiction movies had established certain central UFO themes. By 1952, many believed that for centuries people had been observing disk-like flying objects whose speed and maneuverability far exceeded anything ever constructed on Earth. These flying saucers, believers say, are extraterrestrial spacecraft that are flown by alien beings who are here to observe nuclear testing and other activities that concern them. The U.S. government knows that this is true, but to prevent panic is suppressing evidence and silencing witnesses.[4]

Over the next five decades various elements have entered into the UFO myth, gained and lost strength, and in some cases dropped from sight only to reappear. "Cover-up" theories (which suggest that the government has evidence of the extraterrestrial nature of UFOs but is hiding it from the public) have gained strength over time, and at certain points have eclipsed the debate over the reality of UFOs themselves. In the late 1940s there were rumors that the government had retrieved one or more crashed saucers, perhaps with occupants. These beliefs appeared to weaken but regained strength decades later on the publication of controversial books and the broadcasting of supposed documentary films. In the mid-1950s, people reported contact with extremely attractive, very humanlike, long-haired emissaries from outer space who were said to have brought messages of reassurance in the brave new atomic era. Faith in the reality of our space brethren was not sustained because certain claims (for example, that they hailed from Venus) ran counter to our growing knowledge base, but the past decade or two has been marked by a new wave of contact reports. These reports center on the purported abduction of humans by aliens.

The first reports of human abductions by UFOnauts appeared

as early as 1929. But in October 1966, abduction accounts gained momentum when *Look* magazine published an article on the experiences of Barney and Betty Hill. While traveling in northern New England five years earlier, this couple sighted a UFO. Upon returning home, they discovered that the trip had taken about two hours longer than usual. What might account for this missing time? Under hypnosis, the Hills remembered the UFO's landing and their being taken aboard the alien craft. There, they were studied by aliens who were short and thin with large heads and wraparound eyes. Some parts of the examination had sexual overtones. Under hypnosis, Mrs. Hill was able to recall many details, not all of which were verified by Mr. Hill. Although the hypnotist was not convinced that the events were real, the Hill case became the subject of a successful popular book and later a television movie first aired by CBS in 1976.[5]

The early 1980s saw an increasing number of abduction reports with elements reminiscent of those reported by the Hills. (A 1992 survey led to estimates that as many as 2 percent of the U.S. adult population, or 3.7 million people, believed they had been abducted, but this survey may have been flawed.[6]) Reports came from women, and to a lesser extent men, from many walks of life who thought that something special may have happened to them and who had strange, fragmentary thoughts. Whereas the many reports differed from one another in completeness and details, analysts such as David Jacobs identified certain common themes.[7] These included aliens that bore a strong resemblance to those reported by the Hills, recollections of some sort of gynecological or urological examination, and the big tip-off, "missing time." Here is a composite, fictionalized case based on my reading of many published accounts.[8]

> It is late at night, and Jane Doe has fallen asleep in front of the television. Elsewhere in the house, husband John Doe is sound asleep. Jane awakens to mysterious lights playing about the room, lights reminiscent of the dancing colored lights of police cruisers, tow trucks, and other emergency vehicles. Suddenly, a bright light pierces the wall, and short, slight gray men (approximately 4 feet tall) with large bulging skulls and immense eyes enter the room. Although Jane is afraid and cries for help, John's slumber is undisturbed. Jane is transported to the alien craft, perhaps whisked through the

> wall as if it were of no substance, perhaps beamed aboard a spacecraft much as fictional characters were beamed aboard the starship *Enterprise* in the *Star Trek* series.
>
> Aboard the alien craft, Jane communicates telepathically with several of the small gray beings as well as a tall, thin being that tries to reassure her. Jane undergoes an extensive medical examination that includes special attention to her reproductive system. She may be shown scenes of death and destruction: a nuclear holocaust, a flood, a meteor striking her hometown. Finally she returns home.
>
> Afterward, Jane is sick and experiences headaches and nosebleeds, caused, some speculate, by metal implants intended to help the aliens track her after she has returned to Earth. She may have new, mysterious scars, perhaps as a result of incisions to extract reproductive tissue. For a brief period of time she may show signs of pregnancy.
>
> Later, Jane is reabducted to the spacecraft. This time she is shown babies in incubators and is given one of these babies to hold. Not quite human—and not quite nonhuman—the babies bear a remarkable resemblance to human infants, typically raised in hospitals and orphanages, that are afflicted with the failure-to-thrive syndrome.[9] They are frail, sickly, anemic, withdrawn, and depressed.
>
> At first, Jane's memories of the episode are repressed (kept from conscious awareness) or perhaps covered by "screen" memories (false memories that prevent an accurate recollection of the abduction). Eventually Jane comes into contact with an investigator who puts her under hypnosis. Under hypnotic regression, a process whereby she is asked to imagine herself at the time of the abduction, her story unfolds. The experiences she reports are unpleasant and degrading, and they may have a strong masochistic quality.[10]

The bizarre but horrifying implication of such a story is that we are being visited by members of an alien civilization that, for some unknown reason, is doomed. Human genetic stock is essential to them, either for taking over Earth or a similar planet, or for adapting to conditions on their own world. Because estimates of the number of people who have been abducted are quite large, running to a million or higher,[11] there could be a breeding program to produce human–alien hybrids. The human donors are usually but not invariably women, and for whatever reason, *in utero* breeding takes precedence over test-tube babies, despite what would seem to be the considerable aggravation of keeping track of expectant mothers, ensuring that they do not lose the fetus, and reabducting them to give birth aboard the saucer. The aliens, who can walk through walls and beam living entities hither and yon, have not yet progressed much further than we have in genetic engineering and seem to be running behind us in test-tube incubation.

Thus, over the years, reports of shiny objects in the sky were supplemented with elaborate descriptions of alien spacecraft, descriptions of aliens themselves, and now lurid accounts of alien–human interactions. Perhaps this shift can be explained by habituation, a psychological process whereby we become used to something, and its emotional impact declines: Whether it's about a pay raise, a new car, a sex partner, or a story, initial excitement tends to dwindle over time and we look elsewhere for new, higher levels of stimulation. We have, over the years, become habituated to garden-variety encounters with aliens, so only spectacular claims accompanied by tantalizing detail and intricate theories are enough to grab our attention now.

There are many elements in the UFO story: sightings, contacts, cover-up theories, and so forth. A person who believes that flying disks may be transportation for extraterrestrials does not necessarily believe that its occupants are actively abducting humans, and, of course, it is not necessary to believe in either flying saucers or abductions to harbor suspicions about the U.S. government!

EVIDENCE

What is the evidence that UFOs are extraterrestrial craft? Primarily, eyewitness accounts. Eyewitness reports are corroborated by other eyewitnesses; there are "correlated" radar reports; there are snapshots, movies, and videotapes; there is trace evidence in the form of physical aftereffects; and there is the occasional substance or piece of material said to have come from a UFO itself. All that is lacking after 50 years of sightings, study, and debate is acceptable scientific proof that extraterrestrial spacecraft have visited Earth.

UFO phenomena occur on a haphazard basis, and UFOs themselves cannot be coaxed into a laboratory. There are many areas where field research is necessary or desirable. But unlike UFOlogists, field biologists, for example, have methods for finding their quarry. Certain birds inhabit specific ecological niches, and by visiting these sites, scientists who are patient and properly equipped can get photographic data. While some geographic areas (such as the American Southwest) are reputedly "hotter" than others, UFO sightings are unpredictable events. This means that, unlike the field biologist, the UFOlogist may well be unprepared to make careful observations and recordings. Thus, attempts at documentation are haphazard, rushed affairs.

Since sighting UFOs is hit or miss, one of the chief qualifications is being Johnny-on-the-spot. The accidental UFOlogist is unlikely to have received training as a field scientist. Observations are dependent on procedures as well as on the phenomena themselves, and it is for this reason that scientific papers describe procedures in great detail. In many cases, UFO reports are incomplete and focus only on those elements that strike the reporter as important or interesting. Furthermore, unlike most published scientific papers, UFO reports have not always undergone rigorous peer review. Because so much of the data are based on incomplete reports by untrained observers, their quality is at worst suspect and at best difficult to assess.

After an extensive discussion of such problems, Richard Haines, a leader in perception and aviation psychology, has prepared a set of guidelines to promote comprehensive and accurate reports and thereby improve UFO research.[12] His book presents methods for estimating size and distance, plus diagrams to help observers identify shapes. Haines and other scientifically oriented UFO researchers believe that if we can obtain the same type of information about many cases, we can establish norms and make meaningful comparisons. For example, we could determine the proportion of reports that involve disk-shaped objects, cigar-shaped objects, and so forth. Standardized procedures facilitate quantitative analyses, and such analyses are important for revealing trends. Trends would help us separate consistent themes from random occurrences, and find order in chaos. A large data base, as comprehensive and accurate as possible, could take us a long way toward unraveling UFO mysteries. Several UFO research groups are working diligently to develop high-quality data banks.

Photographs, Movies, and Videotapes

Some of the many photos of UFOs present clear and detailed images, but most give us very little information. They are grainy, blurred, or show bright but ill-defined objects or pinpoints of light. Some photos reveal identifiable terrestrial objects such as pie tins or Frisbees; some pictures are not centered and raise suspicions about off-camera devices supporting the objects in midair; some pictures, computer enhanced or viewed with a magnifying glass, reveal wires or strings holding the objects aloft. Movies have an advantage because they provide multiple images and a running record, but because they tend to be small-format (8 mm rather than 35 mm or larger) the individual frames tend to lack resolution. Videotapes are problematic because they can be electronically dubbed, or even edited pixel by pixel. People who report having been abducted on more than one occasion have set up video recorders to tape their abductions, but these attempts have come to naught.[13]

The problematic nature of photographic evidence is easy to illustrate with an unequivocally real event and a well-known film recording. The event is the assassination of John F. Kennedy in Dallas's Dealey Plaza in November 1963. The film is one taken of the assassination by Abraham Zapruder. An amateur photographer, Zapruder used a small-format, hand-held movie camera with a telephoto lens, a combination that led to grainy, unstable pictures. Over the years there have been multiple copies of this film, copies that differ not only in quality but also (due to editing) in content. Various researchers have sought to apply corrective techniques and enhancements; these have led to the accentuation of some details (for example, the president's head) but the loss of others (for example, landmarks that would indicate the precise location of the motorcade). Even the film's status as the official "timer" of the assassination has been challenged because we do not know whether the spring-driven motor was fully wound when the filming began. With hundreds of eyewitnesses, tape recordings, many still photographs, and home movies, it is still possible to give very discrepant accounts of what happened in Dealey Plaza that late-November day.[14]

This does not mean that good photographic evidence of UFOs is impossible. Ideally, it would be obtained with two or more large-format (35-mm or 70-mm) tripod-mounted synchronized cameras located some distance from each other. The results would allow us to estimate the size and speed of alien aircraft, and the pictures themselves would be almost impossible to fake. Since UFOnauts never announce their impending arrival it is unlikely that we will ever obtain photographic evidence of such high quality.

Artifacts and Traces

Physical evidence tends to be especially elusive. First, there are reports of crashed "saucers" made of amazing materials with remarkable engineering properties and marked with hieroglyphic-like script. Unfortunately, these are unavailable for public inspection, as (according to the legend) this undeniable proof has been whisked away and sequestered by government agencies (events following the crash of secret airplanes such as a stealth fighter prove that the government has the technical means to do this). There are more modest forms of physical evidence, such as pieces of metal that are said to have burned, flaked, or simply fallen off a UFO's hull. In the cases I have followed, such evidence has never been definitively analyzed and often has properties that reveal the possibility of terrestrial manufacture.

Discussions of abductions hint that abductees may be marked for future identification by means of a small metal sphere inserted in the head through the nose. In some cases, we hear, the markers have been expelled as a result of a vigorous sneeze or blowing of the nose, and supposedly unaware of the BB-like object appearing in the handkerchief, the abductee has thrown it away. If there are large numbers of people with unworldly metallic implants, it seems that one or more should have been recovered in the course of normal medical X rays, but here, too, we are waiting.

UFOlogists have reported "trace" evidence in the form of

physical changes in the aftermath of a sighting or abduction. Typically, this includes compressed, scorched, or discolored vegetation; increased levels of radioactivity; and the presence of unexpected chemicals. Whereas some of this evidence is interesting, much of it can be explained without invoking visitors from outer space. Another form of trace evidence consists of stigmata and injuries suffered by people who have come into close proximity to a UFO. Once again we have trouble rejecting prosaic explanations for their appearance. For example, headaches and scars are not compelling evidence of an extraterrestrial visitation unless prosaic alternative causes can be eliminated.

New Knowledge

As mentioned in the discussion of ESP, one form of acceptable evidence of extraterrestrial visitations would be new knowledge—that is, insights that are currently beyond our grasp but that we would recognize as true or be able to verify once they were given to us by the aliens (see chapter 2). In general, people who come into contact with ETs say they forget to ask important questions, have their questions deflected by the aliens (the abductors seem particularly adept at avoiding central issues), or cannot recall the answers. Some of our alleged visitors are good at chatting about political and social issues, at least if you are willing to accept vague allusions, generalities, and platitudes. They are less communicative in the area of science and tend not to venture into technical areas that are far beyond our present knowledge. Aliens that wish to demonstrate their intellectual superiority fail to do so, and once again we are deprived of proof of their existence.

EYEWITNESS ACCOUNTS

After other evidence fails to materialize or to withstand scrutiny we are left with eyewitness accounts. Some argue that the high degree of consistency among different people's accounts provides proof of their veracity. However, corroborating reports should be given full weight only if the accounts are independent of one another. If different reports are based on identifiable and well-circulated science fiction stories of an earlier era, or if one person's account is based on knowledge of another's, then additional witnesses add little to the original report. Moreover, in the case of abductees, other people who are in the immediate vicinity apparently "switch off" at the time of the abduction and cannot give corroborating evidence.[15]

Unfortunately for the hypothesis that UFOnauts hail from outer space, many of the basic elements in the abduction accounts occurred in science fiction stories decades ago. Hilary Evans reports that he was astonished by how many current notions about UFOs and abductions are matched by the science fiction literature of the 1930s. He found "space ships shaped *precisely* like classic flying saucers ... alien abductions in which discreetly dressed ladies are stretched out on tables to be examined medically ... people paralyzed by rays or beamed up in them." He concludes that "there is hardly a concept in the imagined world of today's UFO experiencers that was not prefigured by these early writers.[16] Burned into the collective unconscious, these elements may account for some of the early abduction stories, which in turn serve as models for the later accounts. The widely read book describing the Hill case and, more importantly, the television movie could have provided the inspiration, perhaps unconscious inspiration, for the accounts that followed. Aspects of the Hills' abduction experience are remarkably reminiscent of the movie *Invaders from Mars* and an episode of *The Outer Limits*. As Robert Baker so aptly put it, the UFO abduction scenario is "known to every man, woman, and child newspaper reader or moviegoer in the nation."[17]

Instructive in this regard is a study of hypnotized subjects who were asked to imagine that they had been brought aboard a flying saucer and given a physical exam.[18] They were then instructed to describe what they saw. In the course of the hour-long sessions, subjects produced detailed accounts that were in many ways consistent with "authentic" abduction reports. The researchers had expected that the subjects would require prompting, but the subjects produced the abduction tales easily and naturally. Yet, subjects had been screened to eliminate people who were highly knowledgeable about flying saucers or who might have undergone an actual abduction. This shows that our common cultural heritage provides us with the information we need to concoct an abduction account that fits the classic pattern. It is not necessary to undergo an actual abduction to provide an "accurate" description of the experience.

One account triggers many others. In early 1976, a search of 30 years of UFO literature revealed 50 abduction cases; only two years later, after the movie about the Hills had aired twice on national television, the same researcher found three times as many cases.[19] Spanish researcher Vincente-Juan Ballester Olmos reports that it was not until foreign abduction cases reached the Spanish media that the first claims appeared in Spain.[20] He also showed that, on a worldwide basis, cases tend to cluster following a major media event. Critics of this type of finding counter that well-publicized cases liberate people to discuss their own experiences, but the causal sequence proposed by Ballester Olmos is consistent with what we know about how we learn from and imitate one another, and his findings do not force us to invoke aliens from outer space.

Research and support groups may impose consistency on otherwise conflicting stories. There is strong documentation that when people are in group situations their judgments tend to conform with one another. For example, in a darkened laboratory where there are no bearings, a stationary pinpoint of light will appear to move.[21] (This, by itself, has interesting implications: how many times have you watched a "satellite" move slowly overhead only to realize later that it was a star whose true movement was imperceptible to you?) When individuals are asked to judge the extent of the movement, their estimates vary widely. However, when they make judgments in front of the group, their estimates tend to converge. Subsequent solo judgments tend to fall within the range established by the group, suggesting that once socially imposed, a range becomes "real" to an observer. In the course of discussing UFO sightings or abduction experiences, similar processes may result in the descriptions becoming more similar to one another, and this socially imposed consistency may be misinterpreted as proof.

Perceptual Factors

UFO and abduction reports represent the culmination of a complex series of events. The first of these is the sighting of an object that somehow stands out from those in past experience. It may appear brighter or larger than normal terrestrial or celestial objects, or maybe its behavior makes it distinctive. Perhaps it hovers motionless, proceeds at a terrific rate of speed, or executes turns involving forces that could not be withstood by the human body.

UFOlogists recognize that a significant proportion of UFO reports are misinterpretations of everyday phenomena. Less than a week after Kenneth Arnold reported nine UFOs in the vicinity of Mount Rainier, helpful commentators offered a number of prosaic interpretations: he had seen reflections of his airplane's instruments in the windshield, he had seen the glow of jet fighter plane's exhausts, he had seen flashes of light off nearby airplanes, and so forth. Most creative, was the explanation proposed by ironworks operator Ray Taro, who tried to account for all West Coast UFO reports. As part of this job, Taro boiled down bottlecaps in a furnace, but the temperature was not high enough to melt the inset aluminum disks. Taro theorized that after these little disks were ejected from the furnace's smokestack by a powerful fan, they were carried away by the jet stream, there to glitter in the sun.[22]

Certain celestial objects are truly spectacular. Under some conditions Venus, Mars, and a few stars appear brilliant—bright objects seem larger than life and are not always recognized for what they are. Philip Klass reports that during World War II some U.S. bomber crews in the Pacific theater kept trying to shoot down what they believed was an enemy pursuit plane. Their ammunition was wasted: the "searchlight" plane was a "very bright planet Venus."[23]

At dusk, shortly after the sun has disappeared below the horizon, satellites and weather balloons shine as they pass overhead and reflect the last rays of the sinking sun. Meteorites can also be spectacular, as I discovered while driving near London, Ontario, one evening in the mid-1960s. A meteorite, which came from the east and ultimately landed in upper Wisconsin or Minnesota, lit up the sky. The illumination, as I recall it, was so intense that it triggered a system intended to shutoff the streetlights when the sun rose in the morning. Many viewers called the radio station I was listening to, and it was obvious that the disc jockey was broadcasting from a room with no windows. He initially reported the story as the work of a prankster and made jocular reference to people who believed in flying saucers. His consternation grew as the successive waves of callers confirmed the first reports.

Haines describes how we can misperceive an airplane that is approaching with its landing lights illuminated.[24] At some distance the airplane will be difficult to hear, especially if there is a highway nearby. Aircraft landing lights are very bright and stand out against the normal array of stars and planets. From a distance, the plane's two bright lights, one on each wing, will merge, giving the impression of one large bright object. Furthermore, the luminescent "object" may appear stationary if the plane is approaching head-on, because the frontal mass will increase at a nearly imperceptible rate. What the viewer sees, then, is a bright and seemingly large object hovering in the sky. As the airplane continues its approach, the two lights will eventually separate and the reassuring sounds of the airplane's engines will be heard. However, if the lights are switched off, or if the airplane abruptly changes direction, this identification may never take place. The "UFO" seems to vanish.

Usually, it is difficult for the UFO observer to see details. First, since the UFO is unexpected, the observer has to overcome surprise and then shift his or her attention to it. Second, many sightings do not allow for a good view: the UFO is too small and distant, it is moving at too fast a clip, the observer has only the briefest glimpse. Finally, there may not be enough reference points to gauge its size. Once, at Point Reyes National Seashore, I was treated to an exquisitely detailed radio-controlled model of a World War II fighter—or so I thought, until I saw the pilot move around in the cockpit and realized that the sound of the engine was actually much deeper and throatier than I first detected. The plane was an absolutely spectacularly restored specimen of the real thing! If I had first seen this very expensive toy on the ground, there would have been no such confusion; nearby objects of known size, such as other airplanes, vehicles, hangars, and the like, would have prevented it.

Memory Factors

UFO and abduction reports are based on memory. In some cases the sighting or abduction occurred many years earlier than the report. Human memory is notoriously unreliable. Memories are not stable, like songs etched into a compact disc waiting to be played. They are dynamic and imperfect. They fade over time, are blocked or shaped by newer memories, and are contaminated by the present.[25] Memories, or what we take as memories, may not reflect actual experiences but represent instead unwitting fabrications that occur at the time the "memory" is elicited.

There is growing evidence that people walk around with their heads full of fake memories, including, in some cases, fake memories of abductions by aliens, past-life identities, and ritualistic satanic abuse.[26] While real enough to the person, the memory may have been formed long after the supposed fact. A fake memory can be the result of a two-step process. First, the researcher or therapist coaxes forth an account that fits expectations. Oftentimes, this involves structured interviews that include repeated demands to describe the requisite experiences. In essence, the researcher demands a story that is consistent with his or her theories, and the subject produces it. Then, once this accounting has actually been made, it becomes real to the person who made it, and indistinguishable from memories of events that actually did occur. This general process has been demonstrated again and again in many situations. It is a plausible explanation of hundreds of thousands, perhaps millions, of abduction reports.[27]

During the 1950s there was a spate of studies showing the effects of "social reinforcement" on the rate at which people speak and the kinds of things they say.[28] Simply by nodding our heads in agreement and uttering such remarks as "Yes, I see," "Mmm-hmmm," or paraphrasing what we hear, we encourage other people to continue. If we remain silent or look away, they tend to shift to a different topic of conversation or taper off into silence. Some students who learned about this plotted to shape their instructor's lecture. Students sitting to the instructor's left nodded enthusiastically, asked questions, and in other ways encouraged the instructor; students sitting to the instructor's right sat deadpan and silent. Over time, the instructor's remarks would be directed more and more toward the half of the class that showed interest. Social approval affects both how much we talk and what we say.

Imagine the effects of such processes on abduction accounts. Those whose stories have elements consistent with the researcher's expectations are invited to undergo extensive interviews or join a support group. There, nods and encouragement draw forth further "proof." Statements that are inconsistent with the researcher's preconceptions are met with icy silence. In this manner, the experience becomes more "real" to the reporter (there is social validation of the experience), and the account itself gains consistency with other accounts and is enriched with intricate details.

Nicholas Spanos, Cheryl Burgess, and Melissa Faith Burgess note that reports of UFO abductions, past-life identities, and satanic ritual abuse show certain similarities.[29] In each case, interview procedures are geared toward eliciting reports that fit the therapist's or interviewer's preconceptions. Therefore, studies of past-life identities and satanic ritual abuse may help us understand UFO abduction phenomena. Instructive, then, is a study by Bette Bottoms, Philip Shaver, and Gail Goodman, who found that a large majority of psychotherapists interviewed had no contact with people who reported ritual abuse, but that a small minority reported seeing large numbers of patients who had such experiences. This suggests that the latter therapists may play an active role in shaping what their patients say.[30]

The second step, becoming convinced that a fake memory is real, may become particularly likely under hypnosis. Although theorists are not in full accord, many believe that hypnosis involves two processes: an altering of consciousness that results in the suspension of normal, critical thinking, and the hypnotist's swaying of the hypnotized person. The events we remember under hypnosis are a mixture of fact and fantasy; in the absence of other proof, the proportions are just about impossible to determine. The psychiatrist who worked with the Hills noted that whereas hypnosis may be a magical road to truth, it is "truth as it is felt and understood by the patient. The truth is what he believes to be the truth and may or may not be consonant with the nonpersonal truth."[31] Estimates of the percentage of abduction cases that are brought forth under hypnosis hover around 80–90 percent and in some casebooks reach 100 percent.[32]

Are people who report UFOs or abductions fantasy prone, or are they particularly suggestible or susceptible to hypnosis? Although such speculations are perennial and intuitively compelling, researchers thus far have found scant support for them. A 1993 study of people who had undergone UFO experiences did not find them to be more fantasy prone or hypnotizable than people who were in many ways similar but who had not undergone a UFO experience.[33] Similarly, Kenneth Ring found that people who had undergone a UFO experience did not score appreciably higher on a measure of fantasy proneness. These people did not claim to have a particularly vivid imagination, did not daydream a lot, or indulge in particularly rich fantasy lives. Fantasy proneness may play a role in some abduction reports, but it is not an essential prerequisite.[34] Quite possibly, a sufficiently demanding researcher and normal memory processes (coupled, perhaps, with a preexisting belief in the reality of UFOs) are sufficient to account for abduction reports.

Motivational Factors

Hoaxes are responsible for some UFO reports.[35] What might motivate false claims? The most obvious incentive is personal gain. By weaving an interesting story, a person may achieve a certain fame. A clever individual can turn a UFO sighting or abduction report into material gain by going on the speaking circuit or by writing a successful book. In addition to royalties and speaker's fees, unusual experiences bring attention and in some cases sympathy. As Jacques Vallee has noted, there is a cottage industry surrounding UFOs, an industry based on presenting news from the aliens and providing "services" for people who have undergone encounters. The crudest and most speculative theory can be put to commercial gain.[36] Jamie Arndt and Jeff Greenberg add that some abduction stories "come from authors who are often true believers and/or are profiting either through their hypnosis practice or reporting those stories."[37]

Another motive for a hoax may be to play a joke on someone, to demonstrate superiority in a matching of wits. Hough and

Randles describe the Langsford–Loosely hoax, which involved faking an old book of UFO tales consistent with the accounts that circulate today .[38] These tales were then published as if they were a reprint of a book that had originally appeared in 1871. The impact of this book was electric. Confirmation in Victorian times of reports that astound us today! After two years or so it was revealed that the book was a fake. The effect was to ridicule UFO researchers and shift blame to the gullibility of UFO enthusiasts.

Vallee points out that certain UFO and other hoaxes tantalize, charm, and entertain us and are almost impossible to extinguish. He illustrates this with a detailed analysis of the "Philadelphia experiment," which involved the alleged disappearance of a destroyer from the Philadelphia navy docks in 1943.[39] According to legend, the ship was outfitted with secret equipment intended to make it disappear. In the course of testing this equipment some crew members were killed, some went crazy, and at least one was propelled forward in time. The destroyer was real enough, it did carry classified equipment, two sailors did vanish, and the ship itself did disappear and then reappear. But despite the security classification the equipment was low-tech antisubmarine warfare gear, the two sailors went into temporary hiding to avoid being caught for a minor offense, and the ship's disappearance and reappearance came about by a speedy but conventional round-trip to a nearby port. Unlike merchant ships in 1943, navy ships were allowed to take shortcuts through U-boat–infested waters,

After highlighting several characteristics that make hoaxes durable (such as precise details, colorful witnesses, dramatic sequels, the alleged involvement of notable scientists, official secrecy, and media amplification), Vallee offers several countermeasures for those of us who seek to bring rationality into discussions of unusual phenomena. These include (1) disregarding self-described experts, because they usually lack credentials and gain their fame primarily through meetings with one another; (2) disregarding media accounts, because they are likely to be sensationalized and to suppress critical detail; (3) looking for logical flaws in the argument, especially the invalid use of "therefore"; (4) identifying and removing elements that add drama but are irrelevant to the argument; (5) finding and testing independent sources of information; and (6) disregarding claims of secrecy.

James Oberg points out that it is very difficult to address hoaxes because we are all free to speak and because those who would expose frauds may be subject to greater constraints than those who commit them! There is no law preventing me from claiming to have received wisdom from my space brethren, and there are no laws forcing me to give definite proof. But there are libel and defamation laws that might make it risky for you to unmask me. Because of concerns about lawsuits many interesting but marginally believable stories go unchallenged.[40]

Personality Factors

Behavioral scientists enjoy exploring the differences among people. Thus, a natural question for the psychologist is how people who experience UFOs or report abductions differ from everybody else. One prominent hypothesis is that some of the more bizarre UFO reports reflect psychopathology. Certainly, fantasies about aliens and a reported involvement with beings from outer space or other dimensions are among the symptoms that could earn one a heavy-sounding psychiatric diagnosis. Yet studies of UFO sighters and abductees that involve the major standardized psychological tests used for diagnostic purposes have yielded very consistent results. Neither sighters nor abductees show gross psychopathology, or at least no more than one would expect given the prevalence of emotional problems in society at large.[41]

Gregory Little undertook a study of 102 UFOlogists who were listed in *The Encyclopedia of UFOs*.[42] This compendium included biographical sketches and basic "position statements." The largest proportion of UFOlogists (39 percent) believed that UFOs were of extraterrestrial origin, while a significant proportion (30 percent) did not express a primary belief, and a small percentage (7 percent) qualified as "debunkers." Roughly one-third of the group held less than a bachelor's degree, one-third held a bachelor's or master's degree, and one-third held a doctoral degree. Thus, the UFOlogists were highly educated relative to the general population. UFOlogists who believed in the extraterrestrial origin of UFOs were as likely to hold an advanced degree as those who hedged their bets or were openly skeptical. Thus, a weak intellect or a lack of schooling does not seem to explain an advocacy for the extraterrestrial hypothesis.

Troy Zimmer devised a questionnaire to assess college students' beliefs in UFO phenomena.[43] The questionnaire was devised in such a way as to explore three possible explanations for such beliefs. One of these, the "disturbed psyche" explanation, was tested by questions intended to discover if the respondents were psychologically distressed. The "cultural rejection" explanation was explored by assessing respondents' rejection of their own society, as evidenced in cynicism, pessimism, and a view that evil was triumphing over good. Finally, there were measures of the respondents' beliefs in "alternate realities," or their being caught up in the excitement of possible extraterrestrial life.

Results based on 453 college students showed no support for the hypotheses that believers were unhappy individuals given to primitive modes of thought or that they were "social marginals" who rejected society and expressed their alienation by flaunting deviant beliefs. But the "alternate reality" hypothesis did receive support. Compared with people who were unsure whether UFOs were alien spaceships, believers were more likely to be open to astrology and the occult. They were more likely to have had a personal sighting and to be members of the science fiction subculture: for example, they liked to talk about UFOs and believed that media accounts made UFOs more real. They also tended to be skeptical of government disclaimers about the reality of UFOs.

The 1993 study by Nicholas Spanos, Patricia Cross, Kirby Dickson, and Susan DuBreuil focused on two groups of subjects who had undergone UFO experiences: those whose experiences were not intense (for example, they saw lights in the sky) and those whose experiences were intense (for example, they reported seeing and communicating with aliens).[44] Members of these two groups did not score as less intelligent or more psychopathological than did members of comparison groups consisting of students or people drawn from the community. UFO experiences were not associated with some sort of marginal social status: the members of the UFO groups were "solid members of the North American middle class" and almost all were employed, typically in white-collar jobs. Like Zimmer, these researchers found that the groups with UFO experiences were more likely to believe in UFOs and alien life-forms. They conclude that "beliefs in alien visitation and flying saucers serve as templates against which people shape ambiguous external information, diffuse physical sensations, and vivid imaginings into alien encounters that are experienced as real events."[45]

Many sightings and abductions occur at night. At night, even normal, well-adjusted people undergo an altered state of consciousness: they sleep. As we enter sleep or wake up we may have a "waking dream," essentially a hallucination. According to Robert Baker, the characteristics of such waking dreams are either a sense that one is partially or fully paralyzed or, alternatively, that one is leaving one's body. The hallucination itself is likely to be bizarre—a dead relative, a monster, a little gray being with wraparound eyes. Later, such events seem to have been very real. Many waking dreams, suggests Baker, provide a close match to the kinds of experiences that are recounted in abduction reports. They are not demented or abnormal, he notes, but neither are they proof of alien visitation.[46] Consistent with this, Spanos and his associates found that the most elaborate and bizarre reports involved events that had taken place at night. Some of these events were probably dreams that involved UFOs and aliens. Some involved a sense of paralysis, visual and auditory hallucinations, and a presence that was felt if not seen. Understandably, people who underwent such intense experiences characterized their experiences as more negative than did those who simply reported seeing lights.

Kenneth Ring found interesting parallels between near-death experiences (NDEs) and UFO experiences (UFOEs).[47] Reports of near-death experiences are provided by people who have come close to death or who have technically died (by one criterion or another) but are then brought back to life. Many such people give interesting descriptions of their experiences hovering on the border between here and the great beyond. Themes include floating above their bodies and looking down at themselves (perhaps cut open on an operating table or crumpled in a wrecked car), entering a long tunnel, speeding toward a being of light, and being greeted by dead relatives. The episode ends after someone, or something, concludes that the person should be sent back to the world of the living. Despite prominent superficial differences between NDEs and UFOEs (for example, the being of light is kind and considerate while the alien abductors tend to be cold and detached), there were intriguing similarities: a mysterious journey, strange beings, rebirth. Ring undertook a systematic study to help understand those who have had NDEs or UFOEs.

In this study two groups of "experiencers" had undergone either an NDE (74 cases) or a UFOE (97 cases). There were another two groups of nonexperiencers: persons who were interested in NDEs but who had never had one themselves (54 cases) and persons who were interested in UFOs but who had had no significant UFO-related experiences (39 cases). Subjects in each group completed a variety of questionnaires and psychological tests.

The UFOE (and to a lesser extent the NDE) subjects were sensitive to "alternate realities": for example, sensing that a nonphysical being was present, seeing little people, or otherwise tuning into possibilities that either defy or extend beyond our normal physical world. In addition, the experiencers, especially the UFO experiencers, were more likely to report childhood psychic events such as an out-of-body experience or a correct prediction of the future. Consistent also with earlier findings were their reports of having seen or sensed some sort of alien or nonphysical being when aroused from sleep.

Ring's "home-environment inventory" sought to establish traumatic or negative childhood events. Here, too, the two groups of experiencers scored higher than their nonexperiencing counterparts. This is not to say that all experiencers were neglected or abused as children; it is to say that, compared with nonexperiencers, both the UFOE and NDE subjects were more likely to report childhood abuse and trauma. Responses to yet another test suggested that the experiencers had a greater tendency toward psychological dissociation, that is to separate themselves from the here and now. Dissociation is reflected in "blanking out," a sense that one is absent from one's body, a feeling that whatever is going on isn't really happening to oneself. Corroborating findings have been presented by Susan Powers, who found dissociative symptoms among 70 percent of the subjects who reported an abduction by aliens but among only 10 percent of the subjects who had merely seen flying saucers.[48]

On the basis of such findings, Ring proposes that certain sequences of events can steer a person toward becoming "an encounter-prone personality." Children who are subjected to abuse use dissociation as a psychological defense; they learn to protect themselves by tuning out conditions that could traumatize them. (Powers found dissociative symptoms in 100 percent of a group that reported having been abused as children.) In the course of redirecting their attention from the painful external reality they focus on inner, alternative worlds. In later life, when such people are exposed to a UFO or near-death experience their past experiences with these "nonordinary realities" make them able to dissociate, or slip "into that state of consciousness, which, like a special lens, affords a glimpse of those remarkable occurrences. As a result, they are likely to see and register what other persons may remain oblivious to."[49]

Michael Persinger and his associates suggest that electrical activities in that region of the brain known as the temporal lobe give rise to the "visitor experience," that is, a "conviction that one has been exposed to a real, cognizant entity who originates from cosmic sources."[50] He argues that the subjective ("phenomenological") aspects of these experiences tie in nicely with what we know about temporal lobe functioning, and that people with personality profiles characteristic of temporal lobe lability are prone to these types of experiences. What sorts of external conditions might trigger this experience?

In a series of at least 15 studies, Persinger's research team found that the release of seismic energy (tectonic strain) within a geographic region was associated with an increased number of UFO reports.[51] The relationship between the dissipation of strain energy and UFO reports was clearest in data organized by six-month intervals over a period of decades. For example, there have been many UFO reports from New Mexico, and one analysis based on data from 1940 to 1980 yielded a rank-order correlation (rho) of 0.74 between the number of earthquakes per degree of latitude–longitude units and the numbers of luminous phenomena reported within each unit. This led to a more refined study of all UFO reports in the New Mexico and Colorado region collected by the Center for UFO Studies over a 32-year period. This, like other studies that the authors conducted at different times and in different parts of the United States, also revealed strong positive correlations between seismic events and reports of unidentified luminous phenomena. Additionally, these studies found that the nature of the reports (movement, rotation, colors, etc.) varied in a consistent manner as a function of the reporter's distance from the earthquake's epicenter.

Ring recognizes the significance of Persinger's work and suggests that encounter-prone personalities—both NDEs and UFOEs—have unusually active temporal lobes. Preexisting conditions and beliefs coupled with temporal lobe stimulation (brought about by medical conditions or surgical procedures for NDEs and by tectonic strain, high-tension lines, or other energy sources for UFOEs) combine to produce the near-death or UFO experience. Unfortunately for the simplicity of the theory, it is not entirely clear how tectonic strain triggers temporal lobe activity, and not all investigators have confirmed that the brain waves of people who report UFO or abduction experiences differ from those of everybody else.

What, then, can we say about people who give abduction reports? No one description fits everyone, but some trends are emerging. There is no evidence that abductees are unintelligent or mentally ill, and it has yet to be shown that they are easily hypnotized and fantasy prone. Many researchers have found that abductees believe in alternative realities, beliefs that may be strengthened by dissociative tendencies, perhaps in some cases resulting from child abuse. The belief in alternative realities may make it relatively easy to respond to a therapist or an interviewer who coaxes forth an abduction account, whether or not hypnosis is involved. Once this occurs, the pre-existing belief system helps make the episode seem real. Joining support groups of other people who claim to have had similar experiences reinforces the belief in UFOs and a sense that the experience was real. An admission that one was mistaken about the abduction memory would be conceding an inability to distinguish between fantasy and reality.

People who recount such unusual incidences believe them, and the episode, whatever it is, is of profound significance to them. This means that psychologists, psychiatrists, and other mental health professionals should take such reports seriously and attempt to help people who believe that they have been abductees gain an intellectual and emotional understanding of what has occurred. Whether a person has been abducted by aliens or has interpreted his or her problems in terms of abduction phenomena, members of the helping professions should be willing to listen.

Jacques Vallee once wrote that as a result of science's abdication of any area that might be considered New Age, charlatans and hoaxters are given a free hand.[52] Some people who sight unusual objects or who have unusual experiences are afraid to report them, or are ridiculed or dismissed. As a result, many such people are driven to soothsayers and cults. Fortunately, the situation has eased somewhat in recent years, and qualified professionals are maintaining ongoing dialogues about the best ways to provide help.

Cultural Factors

Many UFO reports, to say nothing of abduction reports, contain elements that seem to conflict with the idea that UFOnauts are superior beings from a more advanced civilization For years, Vallee has noted certain logical inconsistencies and absurdities in UFO reports.[53] For example, could "they" really be seen hundreds of thousands or even millions of times given the presumed expense and difficulty of interstellar travel? Why should an advanced civilization send so many scouts when a society even as technologically limited as ours could probably learn most of what it needed to know using one or two small reconnaissance satellites? How do they materialize, dematerialize, and in other ways flout the known laws of nature? If they are so advanced, why do so many reports suggest that, at least by our standards, UFOnauts are somewhat dumb, bumbling, and inept? Why is it that despite the flashy conveyances, their scientific methods seem primitive compared with ours? Why, despite their almost endless variety, do UFOnauts almost always bear a fundamental physical resemblance to humans? And why is it that people who say they have encountered extraterrestrial beings or their artifacts describe the UFOs and their occupants in ways that are consonant with their own time and culture?

Such considerations have led many researchers to draw on anthropology, folklore, and religion for at least part of the answer. The general theme here is that our beliefs about UFOs may serve a purpose for our contemporary culture, much as beliefs in elves, fairies, leprechauns, witches, and trolls served purposes in other cultures or at other times. UFOnauts are proper mythical beings for the space age, just as angels were proper mythical beings in biblical times.

Thus, reports of UFOs and their occupants bear many similarities to earlier reports of magical and unusual events, but the present version is adapted to the times. Elaborating on the parallels between "angels and aliens," Keith Thompson argues that such symbolic worlds are "real, vital, and filled with significance *whether or not* any particular UFO was the planet Venus or a Venusian starship."[54]

CONCLUSION

After countless sightings and abduction reports, nobody has produced an artifact of undeniable extraterrestrial origin. Nobody has taken a universally accepted photograph of a UFO, and nobody seems to have inside knowledge that could come only from creatures that are smarter than we are. We do have eyewitness reports, but psychological and social processes shape and influence what we see, remember, and report. Consequently, such reports represent a mixture of what actually happened with imperfections in our perceptual and memory processes; our personality, attitudes, and beliefs; and the influence of our friends, acquaintances, and the culture in which we live. Because of such factors, most UFO reports are compatible with mundane explanations. Unless the evidence changes radically, it is unlikely that the issue will be resolved to everyone's satisfaction. There are too many sets of assumptions, too many procedures, too many rules of evidence, too many examples and counterexamples. We are left, notes Thompson, with stories "that are compelling enough to be reported and also sufficiently absurd to be 'unbelievable' and 'inadmissible.' "[55]

Of course, our preconceptions enter into our receptivity to

reports of flying saucers and abductions. We tend not to like information that clashes with our views of the world or is somehow logically inconsistent. If a claim fits, we tend to accept it; if the claim does not fit, we tend to twist it in such a way as to make it fit, or we tend to discard it. In some cases a discovery or a series of discoveries may be so momentous and undeniable that our conceptual framework changes; perhaps future discoveries will force us to rethink the laws of physics to accommodate faster-than-light travel or extrasensory perception. But in most cases, as William James noted in his 1907 book, *Pragmatics*: "By far the most usual way of handling phenomena so novel that they make for a serious re-arrangement of our preconceptions is to ignore them altogether, or to abuse those who bear witness to them."[56]

It is always wise to be open-minded, but it is also wise to be skeptical of claims that run counter to accumulated knowledge. Chemistry, physics and biology did not arise overnight; they are the result of centuries of cumulative effort. Many of the ideas associated with these fields have been checked and rechecked, evaluated and reevaluated. Certainly there are gaps and errors in our scientific knowledge. No doubt, tomorrow will bring new theories and new findings. Nonetheless, scientific progress has been substantial. Claims that run counter to science must offset a tremendous counterbalance; or, as we frequently hear, "extraordinary claims require extraordinary proof."

Scientists are not immune to an interest in UFOs. In the 1970s, Peter Sturrock sent questionnaires to 2,611 members of the American Astronomical Society. Ultimately, replies were received from 79 percent of the membership. Results showed, not too surprisingly, a wide range of opinions, with older astronomers more reluctant than younger ones to accept exotic explanations. Sixty-two respondents had either personally observed a UFO or had detailed knowledge of a sighting. Two respondents reported something like a searchlight playing on a cloud when there were no clouds in the sky, 11 described disklike objects, and 3 described objects that appeared to emit smaller objects or sparks, and in 2 cases sightings were accompanied by problems with automobile electrical systems. Three respondents provided photographs (two of which could be explained as natural phenomena), and two submitted cartoons. One of these, clipped from the *New Yorker*, showed a primitive warrior reassuring others that an airplane passing overhead was only "swamp gas": the second, hand drawn by a respondent, showed an "unbiased scientist" asking an octopuslike creature with a disembodied head, "Er ... uh ... excuse me, sir, would you be so kind as to tell us what kind of natural terrestrial phenomena you represent?" Many respondents believed that physics, psychology, and sociology would ultimately help unravel the UFO enigma. Over 80 percent of these scientists would like to contribute to solving the UFO riddle, but only 13 percent thought there might be a way for them to do this. Seventy-five percent wanted more high-quality information on the subject.[57]

From the project's inception, SETI scientists have recognized a need to distance themselves from UFO buffs, people who believe that they may have been abducted by aliens, science fiction enthusiasts, and those who embrace phenomena such as astrology, the Bermuda triangle, bigfoot, channeling, the magical powers of pyramids, poltergeist infestation, and dowsing. Their desire is to underscore SETI's scientific nature and to make sure that they are not confused with frauds, wishful thinkers, or quacks. This distancing is essential for them to gain the support of the scientific community and to obtain funding. Of the SETI pioneers interviewed by David Swift, many reported an interest in UFOs—but as meteorites, comets, and other interesting natural phenomena, not as evidence of life from outer space.[58] Wisely, the SETI pioneers have been silent on the issue of abductions. When newspaper reports in Washington D.C. have threatened to undermine Congress's confidence in SETI, more intense public relations efforts have swung into play.

Donald Tarter summarizes all of this with a three-point analogy between "safe SETI science" and "safe sex."[59] Abstinence doesn't work. We will think about extraterrestrial civilizations and how to contact them, whether such thoughts are prudent and responsible or not. Given that we must indulge, SETI scientists should avoid high-risk groups. By this Tarter reaffirms that such scientists must disassociate themselves from UFO buffs, cultists, and believers who are not bound by the rules of science. Even if associating with such groups doesn't lead to muddled thinking, it does lead to a loss of respectability in the scientific community and reduced opportunity for government funding. Finally, Tarter encourages his peers to avoid high-risk practices. By this, he means that it is necessary to follow the assumptions of determinism and empiricism and to conduct naturalistic observations. Steer clear of telepathy, clairvoyance, remote viewing, and other new age procedures he advises.

In my opinion, hundreds if not thousands of researchers who have invested perhaps millions of hours have not found more support for the "extraterrestrial hypothesis" than was available in the summer of 1947. Consequently, we should focus on more promising ways, such as microwave searches, to look for extraterrestrial intelligence. Radio signals that are repeatable and verifiable will allow all of us to share in the truth of that discovery. If our goal is detecting intelligent extraterrestrial life-forms, then we must consider UFO sightings and abduction reports to be false alarms.

FOUR

LIVING SYSTEMS

What will ET be like? Science writer Adrian Berry speaks for many when he states that we can only guess about their nature, and that our guesses will be unproductive.[1] Extrapolating from our experience on Earth to understand extraterrestrial intelligence is an enormous task and one fraught with peril. Since there is no proof that we have ever encountered an alien message or artifact, never mind an actual visitor from afar, our lack of knowledge about life-forms elsewhere in the universe is truly profound. Not only is there an absence of data but in addition, two factors that help us understand other people on Earth will not help us understand beings from elsewhere in the galaxy.

First, all people share a common biological heritage. Despite differences in height, weight, skin color, eye color, facial features, hair distribution, and so forth, people throughout the world are tied together by a membership in the species *Homo sapiens*, and this membership means that in fundamental ways all people are alike. Life-forms that arise and evolve elsewhere in the universe may not fall squarely into one of Earth's two great biological kingdoms (plants and animals), let alone humankind's family, genus, or species.

There is a minority view that humans and extraterrestrials will share some genetic material. One possibility is that Earth was visited by extraterrestrials at some point in the past, and that life on Earth is a result of either an experiment (in that some life-forms were deliberately left behind) or imperfect sanitation or decontamination procedures (in which case humans originated in a garbage pile or dung heap). Another possibility, set forth in the Panspermia hypothesis, is that life is a fundamental property of the universe and has always existed. The seeds of life are propelled throughout the universe by stellar radiation or carried by meteorites and comets.[2] Critics find many problems with this theory, including the immense time requirements for the life-bearing germs to propagate throughout the universe, and the hardiness required to survive both the harsh environments of outer space and the destructive forces encountered on entering a planet's atmosphere.[3] Nonetheless, long after many scientists discounted the Panspermia hypothesis, Fred Hoyle and Chandra Wickramasinghe presented interesting evidence suggesting that some viruses may arrive from outer space.[4]

These investigators analyzed disease trajectories during the 19th century, before the airplane was invented, and during contemporary times, when global air travel is common. They found that during the 19th century certain diseases traveled from point to point at least as rapidly as they do today. Because air travel did not exist, we cannot blame the trajectory on travelers who hopped a plane in London and then infected New Delhi. Furthermore, the progress of some epidemics was not as smooth and uniform as we would expect if they had been dispersed by people or animals fanning outward from a central location. Instead, a strain of influenza that appeared in London might have appeared in remote rural areas before appearing in a population center near London, or it might have spread to other continents before running its course in the British Isles. But diseases do come and go with comets. For example, the $3\frac{1}{2}$ year period between whooping-cough epidemics may be linked to bacteria brought in our direction by the comet Encke.

Hoyle and Wickramasinghe also present evidence that bacteria are space hardy. For example, one strain of *Streptococcus mitus* survived two years' exposure on the surface of the moon. Under some conditions, at least, bacteria survive pressures ranging up to

10 tons per square centimeter, and flash heating up to 600° C. The discovery of viable bacteria inside a nuclear reactor suggests an ability to withstand radiation in space. These are not, note Hoyle and Wickramasinghe, "properties that one would expect to have evolved on Earth, but they are all properties necessary for survival in space."[5]

A recent analysis by Michael Mautner suggests that we will soon have the ability to deliver terrestrial life in the form of tiny organisms to other solar systems. This capability will result from rapid advances in all of the necessary technologies: astrometry, navigation, spacecraft propulsion and braking, microbiology, and methods for guiding evolution. The microbes chosen for the mission must be capable of surviving launch, transit, and capture; of growing on nutrients and of surviving conditions likely to be encountered on many planets; and of initiating evolution toward higher life-forms. The use of microbes makes for small payloads and enables longer transit times than human travel would require. Upon arrival at the target destination, the microbial payload would be dispersed in a ring around an Earthlike planet orbiting a sunlike star. According to Mautner, we should begin as soon as possible even if we are forced to rely on low technology; in order to make sure that terrestrial life is located elsewhere before our own civilization comes to an end. With the enough tries, the microorganisms could ultimately produce intelligent beings who will in turn disperse throughout the galaxy.[6]

As for our own origins, Christian de Duve believes that in all fairness the possibility of Panspermia remains open.[7] Yet this theory does not explain the origin of life: it implies only that life is likely to be abundant in the universe, not that that life is similar to us. We and extraterrestrials might have common ancestors of sorts, but at best this would be in the same sense that dinosaurs, humans, fungi, and quahogs are alike.

A second advantage that we have in understanding other peoples on Earth is that the boundaries between different cultures are loose and tenuous. The Leif Erikssons, Marco Polos, and Christopher Columbuses of history have established links among the planet's cultures, links that have multiplied and strengthened over generations of invaders, traders, and vacationers. The advent of the printing press provided a new means for sensitizing people to other cultures. In these days of instant global communication, outsiders may learn of a development in another culture well in advance of many people within that culture itself. Certainly there may be a few lingering pockets of people who are effectively sealed off from one another, and there are cultures that, like Japan in the days before Commodore Perry (or, more recently, Albania) have thwarted the flow of people and ideas. But on the whole, people and ideas move freely, and as cultures become more similar to one another there is a firmer basis for understanding. As far as we know, there have been no leaks across cultures located in different parts of the galaxy. Thus, the common biological heritage and cross-cultural communication that tend to homogenize people on Earth will not limit the diversity of ETs and will not ease our attempts to understand them.

We must also confront an almost overwhelming tendency to ascribe our own characteristics to aliens. Although there are as many as 40 million living species on Earth,[8] images of extraterrestrials tend to bear a striking resemblance to people. Reported aliens vary in height (although most are between 3 and 4 feet tall, they range up to about 8 feet), color (pale white, gray, green, and orange are popular), facial configurations (they may or may not have mouths, noses, and wraparound eyes), and general attractiveness (ranging from incredibly ugly to beautiful "space brothers" with flowing blond hair), yet we should be struck less by the variations than by the similarities. They are upright bipeds; they have arms and hands (although like some cartoon characters they have three fingers instead of four); with a few spectacular exceptions they either lack genitals or keep them covered; and their garb when mentioned bears a remarkable resemblance to our contemporary attire, including human space suits (one piece, silvery, rubbery). In other words, most reported aliens could be mimicked by humans in rubber suits, a great convenience for underfunded movie producers and people with limited imagination.

One rather far-fetched possibility is that the kinds of creatures that we might encounter arose from the same basic genetic plan and then evolved pretty much the same way we did. Another possibility is that such images originate solely in the human mind and represent nothing more than the projection of human characteristics. That fantasy and projection affect perceptions of nonterrestrial entities is evident in the "flying saucer" or UFO literature; people who say they have encountered extraterrestrial beings or their artifacts describe the UFOs and their occupants in ways that are consonant with their own time and culture.

It is in the area of behavior that the projection of our own characteristics poses some of the greatest risks. In the process of understanding others' behavior, our own behavior is a convenient place to start. But because members of other species are not human, anthropologists, ethologists, and other scientists who study animal behavior usually warn us not to do so.[9] If we begin by looking at ourselves we could be guilty of anthropomorphism, the fallacy of imputing human or humanlike characteristics to nonhuman organisms such as a dog, a zoo animal, or ET. We are urged to fight this tendency in part because anthropomorphism leads us to confuse self-generated mirror images of ourselves with reality, and in part because such visions often have a strong evaluative component that leads to unscientific thinking and egocentric conclusions.

There has been a softening, in some quarters, of the injunction against anthropomorphizing. Some ethologists, known as cognitive ethologists, are starting to catch up with contemporary learning theorists who eschew black-box or empty-organism models for those that draw heavily on cognitive processes, and with the general public who have understood their pets all along. Specifically, these "new look" ethologists are mustering evidence that animals think, in simple terms, about matters that are important to them. For example, in his book *Animal Minds*, Cornell University ethologist Donald Griffin describes animals as "conscious, mindful creatures with their own point of view," and he attempts to infer, as far as the available evidence permits, "what it is like to be an animal of a particular species under various conditions."[10]

Griffin reviews three types of evidence suggesting that animals think. First, animal behavior is versatile and adaptable. There are variations within species and over time, and many animals are capable of responding to novel situations in ways that suggest something beyond preprogrammed or conditioned responses. Second, in humans, certain brain activity is correlated with conscious thinking. Comparable brain waves have been found in other animals. Third, the communicative behavior of animals suggests something far more sophisticated than groans of pain. A few insects and many birds and mammals are capable of conveying simple thoughts, and some birds and mammals are capable of deceiving each other and members of other species. Griffin notes that "whereas many early ascriptions of human thoughts to animals were highly unjustified ... when applied to the suggestion that animals might think about simple things that are clearly important to them, this charge of anthropomorphism is a conceited claim that only our species is capable of even the simplest conscious thinking."[11]

None of this is meant to imply that there are no differences among species or that in addition to knowing how to beg table scraps your dog has solved the problem of cold fusion. The point is that we should not automatically exclude ourselves and our own societies as respectable starting points for speculation about extraterrestrials and their civilizations. Yet if we are to generalize from our own experiences, we must follow George Burghardt's distinction between naive and critical anthropomorphism.[12] In critical anthropomorphism, we look to ourselves for leads and hypotheses that are subsequently evaluated in light of observation and experimentation.

POINTS OF DEPARTURE

As science fiction buffs can attest, there are few limits to speculation about extraterrestrial biosocial entities. The creative writer is free to make any assumptions needed to develop an ET that meets the requirements of the story, although as I pointed out in chapter 1, many science fiction writers go to great lengths to violate as few laws of physics as they can. Is a more disciplined approach possible? Are there any rules of thumb that will steer us in promising directions? Are there assumptions and ground rules that will help us develop hypotheses that have at least a chance of being borne out by eventual data? My answer to each of these questions is a tentative yes. A set of bearings already exists to keep us from getting lost in an infinite range of possibilities; it provides a framework that increases the chance that as our imagination soars at least one foot will stay in touch with reality.[13]

Speculation is, by definition, guessing. However, some guesses are more reasonable and compelling than others. We can reduce the proportion of wild guesses (which are based almost entirely on imagination) and increase the proportion of educated guesses (which have some basis in fact) by recognizing (1) known continuities in the universe, (2) the procedures and findings of science, and (3) universal theories of behavior. Whereas these starting points do not make it possible to forecast alien intellects, personalities, and cultures in any detail, they could help us eliminate impossible or fantasy creatures, narrow the range of life-forms and societies that we need to consider, and prompt useful working hypotheses.

Applicability of Science

Terrestrial and extraterrestrial creatures share the same universe. It is constructed of the same atomic building blocks and governed by the same set of natural laws. The philosophical assumptions and the scientific methods that make it possible to study life on Earth should be equally helpful for studying life elsewhere in the universe. The behavioral theories that transcend species, times, and places on Earth offer potentially useful prompts when we speculate about extraterrestrial life.

Consider a group of people confronted with a complicated problem that can be approached in a number of ways. Each person is likely to start with different assumptions and develop a different chain of reasoning. When they compare notes, the group members become confused by each other's ideas and, as a group, make little or no progress. The mystery of life elsewhere in the universe is a complex problem, and as suggested by the voluminous UFO literature, there is no dearth of people who make different assumptions, follow different procedures, and reach different conclusions. Science helps reduce such confusion because it imposes a consistent set of assumptions and rules. Rather than derailing one another, people who accept these assumptions and follow these rules can benefit from the others' progress. The philosophical assumptions that are the foundations for science on Earth provide good starting points for studying life elsewhere. These include empiricism, determinism, and monism.

Philosophers known as solipsists propose that the universe

exists only in our minds and that when we die the universe ceases simultaneously. As scientists, and as creatures that must survive daily life, we embrace the alternate perspective of *empiricism*, that is, the doctrine that a real world exists independent of our knowledge of it, and that best way to know and understand this world is through the evidence that confronts our senses. There are certain benefits from this doctrine: extraterrestrial biosocial entities that we encounter will be real (in the sense that they exist independent of our minds), they will be knowable (in the sense that we can learn about them through our senses), and their existence and nature will be public or verifiable (in that other observers should be able to confirm our observations).

The assumption of *determinism* posits lawful relationships between causes and effects such that if one can identify and manipulate a cause, one can predict or produce a particular effect. All phenomena, including behavioral phenomena, have lawful antecedents. Even the most complex and seemingly whimsical events have identifiable causes: Admittedly, the causal relationship may be difficult to trace, yet the assumption of an orderly flow of cause and effect means that, in theory, extraterrestrial biosocial entities will be understandable. In principle, we should be able to discover what makes them tick.

Also known as the assumption of psychoneural identity, *monism* is the assumption of the identity, or at least the inherent inseparability, of body and mind. This is a critical assumption in scientific psychology: there is no separate life force. The spirit of monism has been captured by Frank Tipler, who points out that "a human being is a purely physical object, a biochemical machine" that is "completely and exhaustively described by the known laws of physics," and that our sense of ourselves is nothing but a specific program being run on a computing machine called the brain."[14]

If we accept the assumptions of empiricism and determinism, many forms of evidence support the principle of psychoneural identity. Brain complexity correlates with cognitive attributes (phylogenetic evidence); maturation of the nervous system is associated with changes in cognitive functioning (developmental evidence); physical trauma, disease, and toxic substances impair mental functioning (clinical evidence); and experimentally induced changes in brain functioning produce changes in thinking and behavior (experimental evidence).[15] The assumption of monism is important in the present context because it spares us from dwelling on purely spiritual or mental entities.

Compared with dualism, monism has the advantage of parsimony, that is, simplicity. Some discussions of UFO abductions even invoke a three-part metaphysical system involving the mind, the physical universe, and some new reality that is somehow different from the first two. A person who has had an abduction experience is assured that it was not all in his head, that there was an element of reality. The skeptic who demands scientific proof is told that whereas the experience wasn't entirely in the abductee's head, it wasn't entirely in the normal physical world of matter and energy, either. In some cases abductees are described as being on the cutting edge of evolution, able to experience, understand, and appreciate realities beyond the grasp of the average astrophysicist, cosmologist, physician, or mathematician. Dualism has befuddled us more than enough, and in my opinion, a tripartite reality is a giant step in the wrong direction.

Science is imperfect and always in the process of "becoming" rather than attaining the status of a finished product. The scientific views espoused by one generation of scientists may differ in significant ways from the views espoused by their successors. Could it be that our current science is too primitive to allow us to think clearly about extraterrestrials? Would we be better off awaiting, or at least envisioning, a bold new science that proves the "reality" of alien abductions, UFO sightings, and other strange or New Age phenomena, and that enables new technologies for faster-than-light and interdimensional travel? Could it be that until some future time when science becomes sufficiently mature, we should operate freely outside the scientific framework?

I believe that if we are to speculate about extraterrestrial intelligence it makes sense to look to the assumptions, procedures, theories, and findings that serve us now, for these provide better guides than do hypothetical alternatives that have no justification at all.

Universal Theories of Behavior

A theory is a set of statements intended to logically and convincingly summarize some aspect of reality. Theories impose order and meaningfulness on observations and ideas that might otherwise seem unrelated. Theories tell us what to expect under conditions that have not yet occurred and allow us to predict the future. They offer plausible accountings or explanations of phenomena and move us toward the realm of understanding.

At this point, we should seek theories that provide useful conceptual frameworks for organizing our thoughts about extraterrestrial intelligence. Such theories should allow us to impose order on our hypotheses about extraterrestrial intelligence and to generate new ones. They should help us frame worthwhile questions, identify fruitful dimensions and categories, and point us toward useful functional principles. They should be consistent with the data we have so far, even though the real proof will have to await the day we can test these theories outside our own back yard. And, we should be drawn to theories that are general in the sense that they apply to many species and cultures and seem to work over long spans of time.

Absolutism, the idea that certain principles are immutable and have the exact same manifestations in all applications, may work in the physical sciences and mathematics, but does not seem to apply in the behavioral and social sciences. Anthropology's doctrine of cultural relativism suggests not only that behavior is culturally conditioned but also that it cannot be understood or properly evaluated outside its own place and time. Yet it is possible to identify some broad principles that underlie pancultural and even panspecies regularities in behavior. In cross-cultural psychology, these principles are known as universalism, as opposed to absolutism and relativism.[16] Universalism is described by John W. Berry and his associates as follows:

> We believe we will eventually discover the underlying psychological processes that are characteristic of our species, *Homo sapiens*, as a whole. Our belief is based on the presence of universals in related disciplines. For example, in biology there are well-established pan-species primary needs (such as eating, drinking, sleeping) even though their fulfillment is achieved in very different ways in different cultures. In sociology there are universal sets of relationships (such as dominance); in linguistics there are universal features of language (such as grammatical rules); and in anthropology, there are universal questions and institutions (such as tool making and

> the family). In psychology it is therefore plausible to proceed on the assumption that we will also uncover universals of human behavior even though (as in these cognate disciplines) there will likely be wide variation in the ways in which these universal processes are developed, displayed, and deployed.*

The point, then, is to look beyond superficialities. For example, although greeting procedures vary widely from culture to culture (handshaking, bowing, kissing, etc.), all cultures have forms of greeting, which fulfill an underlying social requirement.[17] As methods improve and data are amassed we find increasing evidence for the universalist point of view.[18] Universal theories do not solve all our problems, but they may help us achieve focus and give us further hope for understanding either ETs or radically altered versions of ourselves.

LIVING SYSTEMS THEORY

James Grier Miller's Living Systems Theory (LST) offers an interdisciplinary approach that integrates all the biological and social sciences, from cellular biology to international relations.[19] Among the many advantages of LST are its emphasis on the continuity of physical, biological, and social sciences; its insistence that similar or at least analogous descriptive terms and functional principles apply to many biological and social systems; and a degree of generality that encourages the theory's application across different times, places, and species. Because of these qualities it is a useful starting point for organizing and expanding hypotheses about the psychological and social dimensions of life throughout the universe. Astronomers, some of whom have done serious thinking about extraterrestrial intelligence, are used to

*J. W. Berry, Y. H. Poortinga, M. H. Segall, and P. R. Dasen, *Cross-Cultural Psychology* (New York: Cambridge University Press, 1992), p. 4. Copyright 1992; reprinted with permission of Cambridge University Press.

thinking in terms of systems. Consider Emmanuel Davoust's definition of life:[20]

> By *life* I mean life similar to what we already know on Earth. It can be defined by its properties: metabolism, reproduction, adaptation, selection, and evolution.... Some of its properties can be found in inanimate systems, such as the metabolism of fire or the reproduction of crystals. But these are quite simple systems, whereas life is characterized by a high level of organized complexity. This last criterion can give us a more general definition of life: a complex, organized, and open system.*

A living system is real and tangible; each living system is set off from other systems and from its external environment. A living system is not necessarily self-contained, but it is an identifiable whole that is a convenient unit for purposes of analysis. Systems consist of components, and it is possible to specify what is and what is not within the system. The very term "system" suggests patterning, organization, order. Events in one part of the system will have consequences in other parts of the system.

Living Systems

Living systems are open systems. Whereas closed systems are sealed off and self-contained, an open system maintains transactions with its environment. Open systems are separated from their environment by semipermeable boundaries.

This allows the system to bring in matter, energy, and information from the outside. At least for a time these raw materials sustain operations and promote growth. All systems move toward disorganization (more technically, toward a random distribution of molecules, or entropy), but the rule of thumb is that while closed systems run down rapidly, open systems, with their replenishable supplies, can keep going (or, more technically, they can "preserve negative entropy") and even gain organization and complexity.

*Emmanuel Davoust, *The Cosmic Water Hole* (Cambridge, Mass: MIT Press, 1991). Copyright 1991; reprinted with permission of MIT Press.

Incoming matter, energy, or information (input) is somehow processed within the system (as throughput), and some sort of products and wastes are then discharged from the system (as output). The environment reacts to the output; the reaction registers as a special type of input called feedback that changes the system's operations. For example, two types of electromagnetic energy (electric power and radio waves) might enter a radio circuit. Within the system, signals are sorted and amplified, and the output includes heat and sound. Reaction from outside the system—a flick of the power switch, a twist of the volume knob, or manipulation of the tuning dial—constitutes feedback that changes the system's operation.

System Levels

Starting with blue-green bacteria about 3.8 billion years ago, life has evolved "from simple to complex, from small to large, from monoculture to diversity."[21] According to LST, as single-cell organisms led to multicellular organisms and then to increasingly larger and more complex social entities, eight system levels evolved. These are, in ascending order: cells, organs, organisms, groups, organizations, communities, societies, and supranational systems. Each level is considered higher than the preceding one in that it evolved later, is more differentiated and complex, and encompasses the systems at all lower levels. Thus, depending on one's perspective, a system also can be a subsystem (part of a larger system) or a suprasystem (encompassing many other systems). As George Seielstad points out, "Each individual [organism] contains interacting units and is itself a unit in a larger interactive system."[22] Explaining this further, he says: "The whole decomposition of life into ever smaller components resembles a Russian doll, within which is another surrounding yet another, and so on.... But the resemblance is superficial because the hierarchies of life are not distinct from one another ... each is made of units from the level below and helps constitute a part of the level above."[23]

In the course evolving from cell to supranational system, certain structures and functions retain their identity, although specific manifestations change. LST proposes that systems at each level can be described by the same general terms and functional principles, although higher-order systems may have distinctive emergent properties that are not found at lower levels. The discussion in the next three chapters will be organized around three of the eight system levels: organism, society, and supranational system.

An *organism* is defined in its conventional sense, with the exception that unicellular organisms or free-standing cells are excluded since these are accommodated by a separate system level, that of the cell. Organisms are individual, complete, self-contained biological living entities, including fish, reptiles, birds, apes, humans, and all other animals. The level of the organism is the level of the individual "alien," "extraterrestrial," or "ET." Rightly or wrongly, ETs, or at least the kind that we may come into contact with, tend to be described in terms reminiscent of complex, highly evolved animals.

LST includes, at each system level, nonliving materials or artifacts that affect the system's functioning. For example, clothing provides a second skin that protects us from the elements, and wristwatches are "timers" that help us organize the flow of our daily activities and coordinate with one another. For many purposes, the range and capability of the unaided organism is misleading: a person who is legally blind but who has correctable vision may still drive a car, and a person who is not particularly facile with numbers can still make change if the cash register performs the calculation.

Systems at each level have certain indicators that allow us to describe the system and assess its level of functioning. These include observed and inferred properties and processes. At the level of the organism, physical, biomedical, and psychological observations are informative. Thus, measures of height and weight, resting metabolism, blood sugar levels, strength of grip, reaction times, brain wave patterns, visual and auditory acuity, abilities, personality, preferences, and opinions tell us something about terrestrial organisms. A few of these measures should apply to extraterrestrial organisms as well.

A *society* is a relatively large, usually self-sufficient (totipotential) social system, usually identified as a nation-state. In LST, societies are real and include tangible components as well as inferred social patterns and bonds that coordinate people and tie them together. Individuals form groups, organizations, and communities, and in this way are the building blocks of society. For our purposes, the level of the "society" is the level of the extraterrestrial "civilization," or "culture," although these three terms are not synonymous.

A society's physical components include its land, access to a seacoast and to trade routes, natural resources such as arable acreage and minerals, and lakes, rivers, and streams. Artifacts include fortifications, ports, power and telephone lines, cities, towns, and buildings. Societal indicators include measures of system size (land, population, birth rates, death rates), economic performance (debt, productivity, gross domestic product, economic growth), and social welfare (the availability of shelter, health care, and education).

A *supranational system* consists of two or more societies joined together for one or more common purposes and regulated by a decision-making body that operates at a higher level than the bodies governing the individual societies. Several societies may, for example, form a supranational system by joining together for mutual defense and placing their troops under the command of a separate, higher authority. Supranational systems on Earth include the League of Nations, the United Nations, and the European Union. Ronald Bracewell's "Galactic Club"[24]—encompassing hundreds or even thousands of societies—could qualify as a supranational system, at least if its constituent societies defer in some areas to a higher decision-making body. System indicators are the same as at the level of the society, but aggregated across constituent societies. Thus *total* land mass, *total* population, and *total* power production would provide appropriate information about a supranational system.

Subsystems

LST contends that living systems at all levels may be analyzed in terms of matter, energy, and information; it is difficult to get much more basic than this. If we are to make any assumptions about extraterrestrial biosocial entities, the assumptions that they will be composed of matter, will be powered by energy, and will process information are as conservative as any. Living systems comprise 21 specialized subsystems. Some of these process matter–energy, some process information, and two process both matter–energy and information. Rather than attempting an analysis involving all 21 subsystems, I will present, in very broad strokes, characteristics of two generic subsystems that I will later apply to the three selected system levels (organism, society, and supranational system).

Matter–energy–processing subsystems maintain proper spatial relationships among the larger system's component parts and make movement possible. These subsystems also import matter–energy into the system, store it, use it, and dispose of products and wastes. At the level of the human organism, the skeleton helps maintain proper spatial relationships among organs and limbs, and the muscles make metabolism and behavior possible. The mouth is one means for bringing matter–energy into the system, the stomach and liver process and store it, and waste is discharged by several means. At the level of the society, geographic features and human constructions establish spatial relationships, and transportation systems move system components about. Import firms, offshore drilling rigs, fishing boats, and solar panels bring matter–energy into the system. Warehouses store it, manufacturers and processors convert it into useful products, and high tension wires, pipelines, and trucking firms move it about. Export firms and waste-management companies help dispose of the products.

Information-processing subsystems import, store, process, and distribute information rather than matter–energy. These subsystems allow the system to obtain and analyze data and to make and execute decisions. They also provide markers for time-related events and the sequencing of activities. At the level of the organism, receptors bring information into the system, the brain stores and processes it, and the vocal apparatus produces some of the output. At the level of the society, telescopes, spies, and international wire services bring information into the system; observatories, universities, libraries, and intelligence agencies store and process it. Educational institutions, the mass media, and the Internet distribute information. Societies, like human nervous systems, have their functional equivalents of eyes and ears (foreign correspondents), memory (the Library of Congress), and speech (broadcasting systems).

Although LST draws distinctions among matter, energy, and information, there are many links. For example, some sort of informational command is required to process matter and energy, and matter or energy is involved in the movement of information from one place to another. Moreover, changes in one type of input can affect another type of processing or output. A prediction of drought made by a weather bureau (information) can result in the increased conservation of water in a dam (matter) and reduced hydroelectric power (energy). Or an oil embargo that reduces the flow of petroleum into the society (matter–energy) can prompt a decision (information) that reduces consumption or increases reliance on alternative technologies (matter–energy).

Flows and Pathologies

Subsystems, like systems, are real, tangible, and concrete: they are observable (either directly or indirectly), and the observations yield quantitative measures favored in the physical sciences. Living systems are analyzed by mapping the subsystems and the flow of matter, energy, information, and (in the case of certain higher-order systems), money or money equivalents and personnel. Flows are expressed in the types, amounts, percentages, rates, and lags in matter–energy or information processing.

Systems that don't work right, quit, or collapse fail because of pathologies in matter, energy, or information processing. Deficiencies or excesses of matter–energy or information inputs are one cause of pathology. Insufficient matter–energy inputs can cause people to die of starvation and societies to lose wars. Excessive inputs—too much information to process—cause organisms to develop upset stomachs, headaches, heart disease, and other stress-related illnesses and, at the societal level, lead to legislative gridlock and governmental paralysis. Other pathologies result from abnormalities in the internal processing of matter–energy and information. Thus, organisms can suffer from diabetes and schizophrenia, and societies damaged by pollution and industrial spies. Other pathologies, such as birth defects and political revolutions, stem from maladaptive information in the system's genetic code.

System pathologies are evident in many hypothetical scenarios where aliens become a threat to Earth. The death of their sun (decreased energy input) "forces" them to invade us, or maladaptive information in their genetic code prompts them to mate with humans. The challenges of interstellar communication are easily captured within the framework of LST. For example, it may take many light-years for a message to travel from one society to another (information lag), or the more advanced society may lay on so much information that the other can't begin to keep up (information overload).

CYBORGS AND ARTIFICIAL INTELLIGENCE

A cyborg consists of a fusion of living and nonliving components that produces an organism with extended capabilities, that is, one that can perform at a level beyond that attainable by the unmodified organism or one that can thrive in an environment where the unaided organism would falter.[25] With our false teeth, artificial hips, hearing aids, lens implants, and the like, we humans are already making progress toward integrating biological parts with manufactured ones, although we have yet to catch up with such fictional heroes as Robocop, the Terminator, and the Six Million Dollar Man. We would not expect, then, for members of an advanced technological society to be limited to genetically determined physical and intellectual capabilities. To equip themselves to survive under harsh conditions or off their planet, they could find various ways of replacing relatively weak biological components or even entire subsystems with highly engineered replacements.

Perhaps as we move into our own future we will depend more and more on nonbiological parts. At a certain point, notes W. I. McLaughlin, our "manufactured" parts will predominate and we will evolve into machines.[26] This would not necessarily mean that humans would become hominid robots, as certain functions would no longer be required. Someone does not need arms and legs, for example, to live in virtual reality playing video games and indulging in cyberpleasures. Whereas initial microchip implants may help us process information in the cerebral cortex, later implants may take over for the more "primitive" regions of the brain that are responsible for emotions and aggression. McLaughlin believes that we are a short-lived species with a life expectancy of 10^6 years, that our decline will occur somewhere between 100 and 100,000 years hence, and that we will be replaced by the intelligent machines we will create. If this is a standard pattern—the replacement of intelligent species with intelligent machines—then SETI may discover intelligent machines rather than organisms. These machines may house the organisms, or emulations of organisms, that created them. Because such machines could be very compact or could bear little physical resemblance to living organisms, they could be in our solar system already but difficult to find or to distinguish from natural phenomena.

Frank Tipler envisions a complete unity of person and machine. This is but a small jump from his position that we are biologically based computer programs. Currently, our silicon computers cannot process information in the same volume and at the same rate as our brains. But computer technology is rapidly advancing and should catch up within the next three or four decades. Tipler believes that if the computer can duplicate in every detail the programs that run in your neural computer, then in effect it becomes you. (Some philosophers would object to the assumption that an emulation of a person, no matter how perfect, is the same as that person.) If Tipler is right, then perhaps powerful transmitters and receivers (equipped with modems) would make it possible to "travel" from one civilization to another at the speed of light! It might take an incredibly long time for such a transmission to be completed, and we may wonder what it would be like being one-third on Earth, one-third in transit, and one-third in another solar system—but if you live forever inside a computer, then time is something you have.

Setting aside the existential aspect of life within computers, our first confirmation of extraterrestrial intelligence may come from a computer or some form of artificial life. One possibility is detecting a computer that announces its presence and responds to inquiries. The device itself could be on the planet or space station where it was built; it could be a stationary outpost; or it could be a wandering probe. The civilization that built it could be long gone. There are many possible reasons for constructing such devices: as navigational aids, as memorials to one's culture, as robotic explorers. Finding some form of artificial intelligence may be more likely than finding living organisms because artificial intelligence can travel anywhere and survive indefinitely. Thus, signs of artificial intelligence could be sent to spectacular locations throughout the galaxy and, by piquing the curiosity of radio astronomers, serve as a beacon to the planet of origin.

While discarding the possibility of encountering a ghost, spirit, or mental projection, we can at least entertain the possibility of encountering an "apparition" (a projected image) that closely resembles a living organism but lacks physical substance. Conceivably, advanced civilizations could contrive ways to project holographic images, accompanied by remote sensors to gauge viewers' reactions. It could be possible to tailor the image to generate positive reactions from the intended viewers and, if the "locals" prove to be unfriendly, to beat a hasty retreat.

RESEARCH STRATEGIES

Behavioral scientists, like astronomers, physicists, and biologists, prefer data to speculation, and facts to theories. There is, therefore, a certain awkwardness when the discussion turns to extraterrestrial intelligence. How can we understand advanced extraterrestrial life-forms when we are not even sure they exist? Of course, we cannot actually study extraterrestrial organisms, societies, and supranational systems right now. Although some guesses may prove to be more informed than others, they are still guesses.

Nonetheless, even as science can guide our thinking about extraterrestrial intelligence, so it can help us confirm at least a few hypotheses about their existence and nature. Some research strategies can be applied right now to help us estimate what is and what is not likely elsewhere in the universe. Furthermore, if and when detection occurs, the basic tools of science will continue to serve us well, even though we may have to refine specific procedures and measures.

Analysis of Samples

Exobiologists, who study life elsewhere in the solar system and in the universe, can develop the case for extraterrestrial life through the analysis of samples. Meteorites that do not burn up can be chemically analyzed, and samples of cosmic dust can be obtained from airplanes.[27] At least 74 amino acids, the primary building blocks of life, have been found in carbonaceous chondrite meteorites (at least 55 of these amino acids do not occur on Earth), and complex organic molecules have been found in cosmic clouds.[28] Extraterrestrials or their artifacts that are discovered on Earth would be highly informative, but there are no authenticated examples. Extraterrestrials, on the other hand, could find human artifacts in extraterrestrial settings, for example, the *Mariner* or *Viking* probes or articles left by *Apollo* astronauts on the moon.

Remote Observation

The use of telescopes and other instruments for remote observation is the primary method for locating extraterrestrial societies and, once they have been located, will be the primary method for learning about their nature. Earth-bound optical telescopes lack the sensitivity and resolution for direct visual observation of any but the largest constructions and most flamboyant uses of energy, but spectral analyses can reveal the distribution of elements elsewhere in the galaxy. Infrared sensors are useful for detecting societies that use immense quantities of energy. If an intelligent probe were sent in our direction, radar would help us locate it.[29] Through analyzing an electromagnetic "signature," a powerful telescope, or an array of telescopes, might help us infer an invisible planet's size, rotation velocity, and orbit. We might also discover a craft on interstellar travel through the signature of its propulsion system and the "skid marks" that appear in the course of braking, that is, stopping the ship.[30]

Of course, if we enter into active communication, the amount and type of information that ET chooses to share will be quite revealing. We may be able to glean whether they have had experience with other societies like ours, whether they are reclusive or interested in dialogues with other civilizations, and what kinds of information they consider important to communicate. If we enter into a dialogue, even simple answers to simple questions would be tremendously informative. Normally dreary statistics—the age of the civilization, its land area, climate, natural resources, population size and growth, life expectancy, power consumption, and the like—would rivet our attention. At least at first we would dwell on every detail of their history, law, custom, and etiquette.

Even without the ability to decode transmissions, an understanding of the flow of information among different extraterrestrial societies would tell us something about the supranational system. For example, there are higher rates of information flow within than between systems, so an analysis of who talks to whom might give us ideas about trading partners and alliances.

Although it seems doubtful that tomorrow's behavioral scientists will ever visit extraterrestrial cultures, they still might be able to conduct field research of a sort with remote sensors. Perhaps it will be possible to study alien worlds through virtual reality on Earth. The studies could progress at a snail's pace, since the information would have to travel light-years unless it was extracted from a data bank (an emulation of the society, perhaps) in a nearby probe or worldship. At some point, a multigenerational study may not refer to research participants, but to the researchers!

Another way of conducting a study at a distance is through reference works and the secondary analysis of preexisting data. We hope for an *Encyclopedia Galactica* that will, in effect, become available on our joining the Galactic Club.[31] However, this reference work is likely to be incomplete for two reasons: (1) extraterrestrials may not ask the same questions that we do and hence may not have ready answers for us; and (2) at least at first the encyclopedia will have little to say about life on Earth, and other societies may want information about us. Perhaps Earth's own Human Relations Area Files (HRAF) offer a model for certain chapters in the *Encyclopedia Galactica*. These files organize anthropological data by geographic area and cultural characteristics. Many anthropologists have contributed data over the years. HRAF archives make it possible for researchers to find similarities and differences among cultures and to seek underlying relationships through multivariate analyses. In effect, those who use the HRAF are able to conduct original research (on topics addressable through preestablished data sets) without having to travel to the field, a considerable advantage in the present case.

Simulations

Simulations involve creating or at least capturing some aspects of nature and then assessing their effects and behavior. The most powerful form of simulation is the experiment. Here the researcher actually creates the conditions of interest and then measures their effects under controlled conditions. It is not essential to replicate nature in its entirety (although it would be helpful if we could do so), just certain key conditions. Thus, to learn more about the origins of life, we can study chemical changes after we discharge electricity into a mixture of chemicals in a special pressurized container. We are not using real lightning, the chemicals are not an actual primordial soup, and the gasses that pressurize the container are not a true planetary atmosphere. Yet, we believe, we have captured the essentials of nature. Similarly, we cannot study viruses and bacteria in space to ascertain if they can survive interstellar voyages. In laboratory settings, however, we can subject them to radiation, temperature extremes, and other conditions in an attempt to learn something about their hardiness.

A second form of simulation is the analog. Here we study situations that approximate the situations of interest. For example, because we are relatively short of information on how people will adapt to extended space missions that involve large, culturally diverse crews, we study people in spaceflight-analogous environments.[32] These environments include nuclear submarines, Antarctic bases, and other settings that involve elements of isolation, confinement, deprivation, danger, and risk.

Analogs play a crucial role for SETI forecasters, including those of us who would like to anticipate human responses to the detection of an electromagnetically active civilization. As we shall see, at several points in history large segments of the world's population believed there was intelligent life elsewhere in the universe, and their reactions may be instructive. When we turn to the long-term effects of discovery, we will also review what history tells us about the meetings between very different cultures.

Computer models provide yet more simulations of use. The researcher begins with data or assumptions that can be expressed quantitatively. These are entered into a computer that has been programmed to draw a logical conclusion. Computer simulation applies to all system levels. Models intended to help us understand the formation and functioning of extraterrestrial systems include Martyn Fogg's models of the number of habitable planets and the expansion of supranational systems[33] and Peter M. Molton's computer-generated tests of hypotheses on extraterrestrial life.[34] As we shall see, certain abstract computer models that were designed to help us understand relations among nations on Earth may be of even wider applicability.

In *The Last Frontier*, Karl Guthke suggests that high-quality science fiction provides a useful simulation of possible worlds and contact scenarios. My own experience is that the best science fiction writers make meaningful contributions to discussions of extraterrestrial intelligence. Such writers know science as well as fiction and, even more important, know where the science ends and the fiction begins. As Guthke points out, good science fiction writers help us understand the implications of scientific findings. They deal intellectually and artistically with the philosophical and practical issues of scientific advance, and in doing so they may explore greater themes than those in the "realistic" (non–science fiction) literature.[35] Steven Dick argues that the 20th century has been characterized by a high degree of interplay between science and science fiction, and that many scientists have been influenced by science fiction discussions of extraterrestrial life.[36]

CONCLUSION

In our quest to understand extraterrestrial organisms and societies it makes sense to begin with assumptions and procedures that already serve us well. Although any discussion of extraterrestrial intelligence must have a speculative component, empiricism, determinism, and monism are useful touchstones. Moreover, even as the laws of physics and nature extend beyond Earth at some fundamental levels, so may basic laws of individual and social behavior. Variability of expression, however, makes psychological and social universals difficult to discern.

James Grier Miller's Living Systems Theory provides a useful framework for organizing our discussion. One of this theory's attractions is its heavy emphasis on the continuities among physical, biological, and social sciences. LST chooses building blocks—matter, energy, and information—that remain plausible when the discussion turns to extraterrestrial life-forms and civilizations. Above all, LST is a theory of enormous scope and generality. It is intended to cut across species, cultures, and historical epochs.

In the next three chapters we will consider hypotheses about the processing of matter–energy and information at the levels of the organism, society, and supranational system. The hypotheses we will explore represent but a small range of speculation on extraterrestrial intelligence, as we will not dwell on hypotheses based on alternative universes, psychic communication, and the like.

Many of the hypotheses advanced by astronomers, exobiologists, and other SETI scientists are fully consistent with LST and are occasionally expressed in terms reminiscent of LST terminology. It is not coincidental that Davoust defines life as a complex, organized, open system, and that Seielstad describes subsystems, systems, and suprasystems as being in some ways like a set of nested Russian dolls. Many SETI scientists think like living-systems theorists: in terms of energy, matter, information; in terms of microcosm and macrocosm, in terms of subsystem, system, and suprasystem.

Astronomers and exobiologists are sensitive to continuities among cells, organs, and organisms, but it is less clear that they share LST's recognition of continuities among individual organisms and successively larger social systems. Yet, as we shall see, it is the dynamics of these larger systems, especially the supranational system, that may be important for understanding whether or not contact is likely to occur and its implications for humankind.

We concluded with a review of promising research methods: analysis of samples, experimentation, remote observation, and simulations. Of course, none of these methods is perfect. Samples can be contaminated. Studies from afar are not as good as firsthand experience. The primordial soup we cook up in the laboratory is not exactly the same as the soup we might find in the oceans of a distant planet. Contact between extraterrestrials and humans will not be the same as contact between Europeans and Native Americans. The researcher who develops computer models may make faulty assumptions, overlook critical variables, or use inappropriate formulas. Nonetheless, these same procedures with these same flaws have helped us learn quite a bit about life on Earth. There is no reason they cannot help inform us about life everywhere else.

FIVE

ORGANISMS

Astronomer Frank Drake suggests that at twilight, from a distance, ETs might be mistaken for humans. Their heads will be on the top of their bodies, their eyes not too far from their mouths, and they will walk on two feet, but because it makes a better design, they will have four arms instead of two.[1] A major distinction between this description and thousands of others is that this one comes from a noted scientist who has thought about extraterrestrial intelligence for many years. Of course, he offers it as speculation, not fact, and it is meant to be suggestive of the kind of being that a radio telescope search might detect. With reputation and funding at stake, many scientists are reluctant to speculate about the nature of extraterrestrial intelligence and, when pressed to do so, offer vague generalities, steering clear of details such as size, shape, or color.

Is Drake's guess better than anyone else's, one offered, say, by a recent humanities graduate trying to break into science fiction? Scientists do have strategies for narrowing the range of possible intelligent life-forms that we might detect. One strategy is to draw on broad principles of physics and biology and on the generic requirements for life. Whereas we may not be able to prove that a given life-form exists, fields such as physics, chemistry, and evolution suggest that some life-forms are implausible or cannot exist at all. For example, many of the oversize humanlike beings of science fiction are impossibilities of structural engineering and could not survive under the conditions their creators describe.[2]

Another strategy rests on the view that nature has only so many ways of doing things and that biological structures (which successfully perform the same functions repeatedly) will evolve again and again. Biological structures that have evolved repeatedly on Earth seem to be the more likely candidates than structures that have evolved only once or twice or have never been seen at all. A third strategy is to identify the physical, intellectual, and motivational requirements for any organisms that would undertake interstellar broadcasting or engage in a radio telescope search.

Physicist David G. Blair is among those who are willing to speculate about the kind of extraterrestrials a radio telescope search might uncover.[3] In his view, ET is curious and will have a need to explore and understand the universe. Unless ET is an exploratory animal, there will be no motivation to make contact, since the vastness of space makes many other motivations, like power and greed, meaningless.

Because we cannot hear from a civilization that has yet to reach our level of technology, and because it is so unlikely that we will hear from one that is exactly our age, ET will come from a very old civilization. Over the course of this civilization's long life span, ET will have discovered other radio-using civilizations, and many different extraterrestrial civilizations will form a vast communications network.

Logic and mathematics will be familiar to ET, because these are universal and crucial for the development of high technology. ET will be "ecological," as even advanced societies must be mindful of such problems as energy depletion and pollution. Finally, as a member of an ancient civilization, ET will be patient.

To this daring list let us add two more characteristics. As we shall see, technology and prosperity go hand in hand. Since ET will come from a very advanced civilization, we will see ET as prosperous, even as humans of five, two, or even one century ago would see us as very rich. Second, we will review strong convergent evidence that very old societies are likely to be both democratic and peaceful. Thus, whereas ET could be almost anything, radio telescope searches may be biased toward detecting extraterrestrials that are inquisitive, logical, technologically advanced, peaceful, prosperous, ecology minded, and already experienced in interstellar affairs.

Will ET qualify as a "person"? This is, of course, a philosophical question, and the answer depends on the criteria we establish. Consciousness, a sense of "self" (a belief that one is distinct from the environment and from other persons, coupled with a sense of coherence and stability over time), and the belief that one is a center of action and is free to make choices help define a "person."[4] From there, one can become more demanding, insisting, for example, that persons are defined by other people's reactions to them, or must be able to ponder their own existence. If extraterrestrials are identifiable wholes, intelligent, conscious, and responsible for their own activities, then they are persons. Furthermore, we are better off giving undue attention and respect to a nonperson than treating a conscious and willful entity in a disrespectful manner.

MATTER–ENERGY PROCESSING

Living organisms are real, have substance, and may be described (at least in part) in purely physical terms. ET will be assembled from chemicals drawn from a set that is already known to us and can be described in such straightforward terms as age, size, and weight. Physical structures will demarcate ET's boundaries, hold ET together, impart shape and form, and maintain proper relationships among internal parts. ET will have energy requirements, and there will be identifiable metabolic and physiological functions, even if these differ radically from our own. ET will be able to move and act.

Substance, Size, and Structure

We know the chemicals that provide the basic building blocks for all terrestrial living systems. We know that these chemicals are abundant in our galaxy and we know, in general terms, how they were put together to form life here. We expect ET to share these basic chemical building blocks.[5] Although there may be some exotic possibilities, carbon and silicon are the most promising as candidates for ET's basic stuff. Both are soluble and both can form long chains, but in the final analysis, silicon cannot compete with carbon.[6] First, on planets similar to ours, silicon lacks solubility. Second, whereas both carbon and silicon have the capacity to form bonds, silicon lacks the capacity to form the all-critical "intermediate-strength chemical links firm enough to keep a human being from collapsing into a puddle of warm jelly but weak enough to permit muscles to flex without preliminary treatment by a blowtorch."[7] On Earth, silicon is far more abundant than carbon (in the proportion of about 600:1) and yet it is carbon-based life-forms that prevail.[8]

Earth has an abundance of tiny creatures. According to one estimate, there are 850,000 animal species under 10 inches in length, and 1,500 over 10 inches. Of the latter, only 1 in 150 is over 100 inches in length. A fly would fall at the midpoint of a list ranging from a bacterium to a blue whale.[9] However, Earth may be atypical in that really large life-forms, such as dinosaurs, became extinct 65 million years ago.

John Barrow points out that small creatures—about one-quarter our size—would lack the strength to split rocks and deform metals. Because there is a limit on how small a flame can get, truly tiny creatures couldn't get close enough to fuel and manage fires.[10] Consequently, it is unlikely that small creatures could develop the technology for an interstellar search. Even if they had the physical strength to develop high technology, tiny creatures would probably lack the neural differentiation and complexity required for developing an electromagnetically active society. In general, intelligent creatures need substantial information-processing subsystems. These, in turn, make strong demands on metabolism and large metabolic systems require significant frameworks to keep them in place. Truly massive creatures require tremendous supporting frameworks, and huge quantities of food. Such organisms may have difficulty surviving over the aeons because as they reproduced and multiplied there would be a dwindling availability of food.

Upper limits on size might be relaxed on a planet that is

covered with water or some other supportive medium, but creatures brought up in the sea or a thick gas atmosphere may lack two prerequisites for interstellar broadcasting. They are unlikely to have had the opportunity to experiment with electricity and are unlikely to have their curiosity piqued by a clear view of the heavens. Very large creatures could have information processing constraints stemming from the sheer distance that their nerve impulses must travel.[11] When we consider the distribution of sizes of life-forms on Earth, those species that we suspect of having the greatest intelligence (humans, apes, dolphins) tend to fall within a relatively narrow range.

Membranes, skins, and exoskeletons mark the boundary between the organism and its environment and (along with endoskeletons) help to hold it together. On Earth, boundaries range from the diaphanous and gelatinous membranes of jellyfish or Portuguese man-of-wars to the thick hides of elephants or crocodiles and the hard shells of clams. The strength of an organism's boundary will depend on the medium in which that organism lives. Some creatures' literally thin skins are acceptable in the supportive and homogenous conditions of the sea but would not survive the sharp obstacles or varying conditions of land.

An organism's skin, or "birthday suit," must be adequate for survival in its natural environment. Yet humans and most likely other technologically advanced creatures wear "second skins" that provide additional protection. Suitable clothing makes it possible for humans to survive where they would otherwise perish. The second skin allows people to withstand temperature extremes, poisonous or otherwise unhealthy atmospheres, and other dangers.

Human experience thus far suggests that deep-sea diving gear, space suits, or other outfits (such as garments for chemical and bacteriological warfare) are imperfect mechanisms for adapting to unusual environments. Even simple tasks become difficult to perform and the wearer tires easily. Long-term human habitation of the moon or Mars may require space suits that weigh half as much as *Apollo*-era space suits and offer twice the mobility. Quite likely, extraterrestrials that have moved into space will have developed protective gear that does not impose such significant handicaps.

Skeletons help arrange and separate internal parts as well as play a critical role in morphology and mobility. Skeletons may be internal (endoskeletons) as in the case of humans, or external (exoskeletons) as in the case of insects. Insectlike creatures are popular in science fiction, but in large organisms exoskeletons do a poor job of organizing and spacing internal parts. Furthermore, exoskeletons could impede progress toward high technology. A bony or horny outer layer is unlikely to allow the degree of dexterity required for manufacturing radios or starships. Dexterity requires not only smoothly working joints but also sensors on the organism's perimeter to provide the level of tactile feedback required for making or using tools. Developing a computer chip while burdened with a full exoskeleton would be like trying to do so while wearing boxing gloves or with fingers shot through with novocaine.

As organisms increase in size, the skeleton increases disproportionately, with the result that very large creatures would be almost all skeleton. It's not simply a question of scaling up with increases in size. If humans weighed 600 pounds or more, the skeleton would have to be redesigned for the human to survive on Earth.[12]

The appropriate size and strength of a skeleton depends also on the organism's ecological niche. Consider a fairly large organism that evolves on a planet approximately the size of our moon, which has a small fraction of Earth's gravity. The skeleton may be sufficient to support the organism's mass on the home planet, but it will not be capable of supporting the mass in the full gravity of Earth. Alighting on Earth, the creature's skeleton might collapse. Perhaps the landing vehicle used in the *Apollo* moon program provides a useful analogy. In one of many efforts to keep the weight to a minimum, the ladder used to exit and enter the lunar module was extremely lightweight and would qualify as flimsy by terrestrial standards. This ladder, used successfully by astronauts on the moon, would have crumbled had it been used by the same astronauts wearing the same gear on Earth.

Bilateral or radial symmetry characterizes terrestrial organisms, from the largest whales to the smallest bacteria, with the exception of a few animals such as shellfish that have a helical, or screwlike, structure. Expectations are that most life-forms elsewhere also will be symmetrical.[13] The implications of radial symmetry have been traced by Joseph Royce, who suggests that a radially organized ET would be able to receive and coordinate input from all directions.[14] Compared with their bilaterally organized counterparts, radially organized creatures would have better integrated nervous systems, a superior ability to maintain balance while scampering around, and a greater capacity to conceptualize an *n*-dimensional world. Thus, perhaps Drake's four-armed ET will be more reminiscent of a starfish or coatrack than of a human! Royce suggests that despite their differences, bilaterally and radially organized creatures should have enough in common that much of what we know about the former will help us understand the latter.

Metabolism and Aging

Certainly, it is difficult to envision life without metabolic processes, but such processes may differ substantially from those of humans. Terrestrial species differ tremendously in their oxygen requirements, the amount of water and food they must consume, and their metabolic rates. Some species such as hummingbirds must consume food almost continuously to meet their caloric requirements, whereas others such as snakes get by with weekly or biweekly feedings, while bears can store immense amounts of energy, turn down their metabolism, and survive in hibernation for extended periods of time. Perhaps organisms that must engage in near-continuous matter–energy processing would have little chance or motivation to develop an interest in scientific exploration.

On Earth, life, or metabolic processes, proceeds at different rates for different organisms. Absolute limits may exist, since events do require some amount of real time (for example, for molecular motion). By human standards, extraterrestrial time scales could be much compressed or expanded. On Earth, smaller animals' metabolic rates are faster than those of larger animals. Size and timing are negatively if imperfectly correlated, so that very large species tend to be somewhat slow-moving.[15]

Terrestrial life spans range up to three or four centuries for certain tortoises and millennia for some trees. Although a few creatures, such as terrapins, outlast humans, we are one of the longest-lived terrestrial animals. The oldest authenticated human age is just under 121 years. We should have little confidence setting upper age limits, however. Some writers have estimated genetically engineered life spans on the order of 3,000 years, or 150 generations.[16] Even more optimistic estimates suggest that, for all intents and purposes, life can be extended indefinitely.[17] A culture where this is so may place a very high premium on safety, since accidents would be the leading (and perhaps only) cause of death.[18]

Immune systems fend off or cure infection, and repair mechanisms lead to the coagulation of blood and the regeneration of cells and tissues. Some of these mechanisms are effective only in a particular ecological niche and at particular times.[19] Throwing organisms together from different locations or time frames raises the possibility that one organism will host viruses or bacteria that are dangerous or lethal to the other. Even meeting your great-grandfather brought forward in time could have deadly effects. He might not withstand the onslaught of modern diseases; nor could we assume that your immune system would provide the same protection as would your grandfather's against the illnesses of the past.[20]

Motion and Motility

Living systems are capable of motion. At the lower levels of our evolutionary scale we find cilia, flagella, and pseudopods; at the higher end we find muscles, arms, legs, and wings. Muscles are necessary for internal processes such as heartbeats, inhalation and exhalation, and the peristaltic movements required for digestion. External manifestations include locomotion and dexterity. Locomotion encompasses any form of movement through the environment, including walking, swimming, and flying.

Terrestrial creatures include some that more or less stay put throughout their lives or at least move very slowly (such as snails) or only when impelled (such as jellyfish). Creatures that are stationary or that are moved hither and yon by winds and tides are unlikely to develop science and cannot gather the materials required to build transmitters or spaceships. Other creatures are highly mobile and dart from place to place (such as hummingbirds) or undertake long migrations (such as certain birds or whales). Such creatures may not stay put long enough to develop the necessary science.

Dexterity is important for making and using tools. Creatures who must use all their limbs for support or balance will not become tinkerers, nor will creatures that can't physically grasp tools. Creatures who lack dexterity or fine motor skills are unlikely to be able to assemble the necessary communications or travel devices.

Reproduction

Reproduction, as many of us think of it, involves the mixing and matching of genes. Chance variation provides the opportunity for natural selection and evolution. The combination of genes that defines the instant of conception sets in place a template for biological organization and growth. Different species follow different reproductive strategies. Animals that are large enough and strong enough to control fires and work metals, and that have the structural and metabolic wherewithal to support a large brain, are not likely to rapidly produce huge numbers of offspring. Rather, they are likely to start reproducing relatively late in life and sequentially give birth to small numbers of offspring that require considerable individual attention and care.[21]

Reproductive processes are affected by technology (e.g., test-tube babies) and by social policies (e.g., enforced birth control to compensate for a slow "turnover" due to very long life spans). As technology advances in our own society, capabilities that were once essential for survival may become irrelevant. Engines, servomechanisms, and levers and gears ease physical burdens. A petite young woman who could not drive a team of oxen can drive a truck that bears hundreds of times the oxen's load. A person who can barely make it up five flights of stairs can, in an elevator, be whisked a hundred floors upward. Most of us do not need a high level of physical fitness to survive past middle age or to do our jobs. Supportive economies, welfare states, legal systems, and medical technology keep us in the gene pool even if we are less than fit by the standards of earlier generations.

Could intelligence lose some of its survival value? Already expert systems assist us with automobile repair, accounting, medical diagnosis, and their own design. In a highly technological society, where intelligent machines could repair, replicate, and improve themselves, there could be very little mental challenge. Intellectual burdens would lighten as professional and technical services were taken over by computers and automated systems. Lacking pressures for increased intelligence, intellectually advanced populations could backslide. Thus, there might be civilizations that are highly advanced in the technical sense but whose citizens, by our standards, have an appallingly low level of intelligence. Self-organizing and self-advancing technology would pick up the slack. On the other hand, intelligence has many manifestations, and it might directed in new directions and take on new forms.

Genetic engineering is another reproductive consideration. With genetic engineering there need be little lag between environmental changes and adjustments in the organism's genetic plan. We could dispense with mutations, assortative mating, winnowing processes to cull out creatures with the wrong traits, and other time wasters. The desired result could be achieved within one generation, or a very limited number of generations. Could a society define standards for strength, intelligence, and beauty and then ensure that everyone met those standards? Could there be a society whose members are "perfect" and completely alike?

For two reasons I do not expect genetic uniformity. First, variability provides survival value for a species as a whole. If everyone is alike then everyone has the same weaknesses as well as the same strengths. A plague or a catastrophic change in climate could wipe out the entire species. If there's variability within the gene pool, some organisms have a chance of surviving the catastrophe and continuing to reproduce. Second, even as there are pressures within societies for members to be in some ways similar (uniformity), so are there pressures for them to be in some ways different (differentiation). Pressures for differentiation should work against efforts to engineer everyone to be genetically alike.

The movie *Jurassic Park* popularized the idea that through genetic engineering, extinct species could be brought back for scientific or other purposes. In this fictional account, biologists harvested DNA from insects preserved in amber. Chemicals were used to isolate the DNA. Since DNA deteriorates over time, genes from cousins like frogs were used to complete the genetic strands. The genetic instructions were implanted in a cell that was in turn placed in an incubator and, some time later—*voilà*—live dinosaurs! *Jurassic Park*'s director, Steven Spielberg, suggested that scientists will be able to follow this general recipe in less than 50 years.[22]

Another important consideration is speciation. Through evolution, species adapt to their ecological niches. Requirements for survival on the home planet and elsewhere may differ, so the mixing and matching of genes, accompanied by natural selection, could nudge homebodies and émigrés along diverging evolutionary paths. Interstellar migration will split humans into innumerable small and reproductively isolated groups, which will lead to rapid physical and social evolution. As Ben Finney and Eric Jones note, "If our descendants do succeed in scattering far and wide through the galaxy, there will be no single future for humanity; indeed, there will not be one humanity."[23] The same should hold true for extraterrestrial spacefaring societies.

INFORMATION PROCESSING

Living systems input, process, and export information. At the level of the organism, information-processing subsystems are responsible for sensation, perception, learning, memory, problem-solving, decision-making, and language. It is an article of faith that ET's intellectual capabilities will be far superior to our own. An expectation that the human brain will triple in the next 900,000 years,[24] coupled with an expectation that their civilization could be a billion years older than ours,[25] hints at the possible magnitude of these differences.

Perception

Thousands of highly sensitive nerve endings are scattered throughout the bodies of "higher" organisms. Called receptors, these are the ports of entry for incoming information. Some receptors, located on or near the surface of the body, respond to events that occur outside the organism. Other receptors (called proprioceptors) are located well inside the body and respond to such events as muscle movements, or to the lack of substances necessary for life or comfort. In some areas of the body (such as the eyes, ears, nose, and mouth) receptors are grouped close together. In other areas (such as on the surface of the back) they are spread far apart. Receptors have the capacity to transform physical events (which cannot be processed by the nervous system) into electrochemical events (which can be so processed). For example, receptors in the eye convert light into impulses that can be processed within the nervous system.

Often we think of humans as having five senses: sight, sound, smell, taste, and touch. (In fact, these "individual" senses rest on a number of different receptors, such as the cones in the eye, which respond to different colors.) Which of the five senses will be available to ET? Will ET have substitute or supplementary senses, perhaps based on infrared energy, electrical fields, or processes that are analogous to radar or sonar? Will they be, for example, like the platypus that can detect prey by means of the very weak electrical fields that surround their quarry? Will certain senses be so highly developed that one can substitute for others?

An extraterrestrial's sensory apparatus might also differ from ours in terms of the sensory ranges to which the receptors respond. The visual spectrum represents but a portion of wavelengths within the range that constitutes light: those that fall between 16- and 32-millionths of an inch. This means, says Richard Haines, that "we are functionally blind to fifty-nine of the sixty octaves of radiation."[26] Other intelligent species may be able to "see" stimuli that fall within parts of the spectrum that are inaccessible to us. They may be able to detect wavelengths that are available to us only through the use of infrared-sensing devices or perhaps even X-ray technology. Most likely, receptors are linked to the local spectrum—if we lived on a planet where there was heat but not light, we would evolve infrared sensors.

Sensory ranges are illustrated by differences in the ability to detect various auditory stimuli. Some vibrations, such as those produced by the motions of the boards within your house, are usually too low to be detected by the human ear, whereas other sounds, such as those produced by the dog whistle, are too high pitched to register on us. Other species may have no difficulty detecting sounds that fall outside the range of human adaptation. Bats are able to transmit very high frequency sounds and navigate by timing the direction and speed of echoes—sort of a living sonar. Elephants and crocodiles communicate great distances by means of very low frequency sounds. This explains certain mysteries, such as why, in the apparent absence of communication, crocodiles rapidly assemble to devour prey. In point of fact the scout has called the others, but the communication is at too low a frequency to be heard by humans.

Another possibility is that ET will have pretty much the same sensory apparatus as we do, but that it will be tuned differently. In other ecological niches, survival may depend on receptors that are much more or much less sensitive than our own. For example, on a blast-furnace planet that was close to its sun and lacked much in the way of a protective atmosphere, receptors might require extraordinary levels of light to activate. For the inhabitants of such a place our high noon could seem like the middle of the night. On a dimly lit planet, creatures may evolve highly sensitive eyes that make it possible to see clearly under conditions that would be too dim for us. Such differences are already evident in comparisons of terrestrial animals that prowl by day with nocturnal animals that prowl by night.

Yet another possibility is that their senses will respond to the same stimuli as our senses do, but that they will not experience them in the same way. This could reflect differences in neural structures. For example, people who have experienced damage to a particular part of the temporal lobe cannot detect faces, although they can identify other people by name, voice, or even clothing. Shown a face, they are at a complete loss, as if their "face detection" centers were missing.[27] Past experience, expectations, and other "soft" factors can also lead to great differences in perception. Because of our experiences with playing cards, we might be "tricked" into seeing a red ace of spades as black. Yet, despite the powerful role of experience, there are remarkable similarities in the ways that people from different cultures view the world. Early studies found substantial differences in how cultures evaluated works of art, but these arose from differences in artistic conventions. When viewers are given actual stimuli rather than graphic renditions to evaluate, cross-cultural differences tend to disappear.[28]

What types of sensory systems might we expect of creatures that would initiate radio communication or begin to fan throughout the universe? It is almost certain that they have receptors that can detect stimuli at great distances. SETI scientists believe that an organism that cannot see the heavens has no reason to think of other planets, much less think that these planets might be inhabited. Sight is the primary candidate. Something like sonar or radar will not work over the required distances. Between the power requirements and the amount of time that would be involved, there is no way that a superbat on Earth could locate Alpha Centauri.

According to a recent analysis by John Barrow, planets with conditions conducive to the evolution of life are also likely to encourage the evolution of color vision. The ability to identify different colors confers many advantages. Color attracts attention, gives warning, makes camouflage possible, and acts as an important emotional stimulus. Color vision is helpful for avoiding danger and for the life-sustaining processes of hunting and gathering.[29]

The chemical senses such as smell and taste would play a relatively minor role in ET's life. The chemical senses do have powerful effects on some species; some creatures respond to subtle variations. But these senses are relatively slow-acting. Moreover, a rapid sequence of stimuli leaves the nose "confused"—parts of scents remain within the nasal passages and take some time to be flushed out. Chemical senses such as smell may be fine for attracting a mate or warning of danger, but they are not helpful in situations requiring a rapid flow of complex information, such as figuring out astronomy or developing a transistor.

Consciousness

Contemplating consciousness in the universe, Joseph Royce hypothesizes eight levels ranging from free, or entropic, energy to "omniscience." The most advanced humans, he says, have achieved the penultimate stage, individuated consciousness. This is "the level of consciousness which sees maximally by means of all of the available ways of knowing" (empiricism, rationalism, metaphorism) but it falls short of omniscience, the ultimate state of wisdom or enlightenment. Royce suggests that one's level of consciousness will depend on one's position on the evolutionary scale.[30]

On Earth, consciousness is not limited to humans. Reminding the reader that other animals are not humans, Carolyn Ristau notes, "It is highly unlikely that given all the continuities between humans and other organisms that humans alone should be aware or conscious, and have thoughts, purposes, beliefs, and desires."[31] This kind of claim is based, in part, on evidence suggesting that nonhuman animals engage in forward planning, are aware of others' intentions, and use symbols.[32]

Humans operate in a presumed "normal state" of consciousness and in states known as altered states of consciousness. These result from physical or chemical actions on the nervous system and include normal cyclical chemical changes within the body, the ingestion of alcohol and drugs, and car accidents or other forms of physical trauma. Could ET's "everyday" state of consciousness be akin to one of our "altered" states of consciousness? Are there organisms that, by storing huge amounts of energy or having a very slow metabolism, are in a state of perpetual sleep? Are there organisms that survive on drug-induced or otherwise induced fantasies that are only occasionally interrupted by reality? Perhaps there are organisms with no sense of time, or with no ability to think ahead. But it is doubtful that we will ever meet them. Creatures that are in perpetual sleep or dream states or that fail to connect with our physical reality or that have no sense of the future are unlikely to make themselves evident to humans or lifeforms on other planets.

If extraterrestrial organisms have entirely different brain structures, the possibilities seem limitless. Studies of the normal cognitive functioning of humans are unlikely to provide us with an adequate base to understand them. We must also turn to comparative studies of other terrestrial species, as their anatomical structures and chemical processes may give us a better idea of what to expect in ET.

Ubiquity of Intelligence

"Intelligence" refers to the ability to identify stimuli, perceive relationships, communicate with others, profit from experience, and cope with change. Intelligent organisms have the capacity to understand the world about them. They can reason and they can choose and pursue worthwhile goals.

Despite commonsense notions about intelligence (or perhaps because of them), extraterrestrial intelligence may not be all that easy to recognize, and we should have no illusions that we will be able to measure their intelligence with our tests.[33] For interstellar comparisons, it may be safest to follow LST's lead and think in terms of the amount and rate of information acquisition, storage, and processing. For purposes of reference, human memory is capable of storing about 10^{15} bits. In terms of processing rates, measured in operations called "flops," the human brain is capable of 10 teraflops, that is, about 10 trillion operations per second. By 2030, desktop computers should be able to match this performance.[34] If ET lives forever, then ET's memory must be very large or very selective.

SETI proponents believe that intelligence is common in the universe; otherwise there would be very little chance of a successful search. How widespread is intelligence on Earth? Recent studies of animal behavior suggest that many animals are more intelligent than most people recognize. All sorts of animals are capable of identifying and relating stimuli, setting goals, and taking appropriate action.[35] For example, humans are not the only tool users. Members of many species extend their efforts by means of simple technology. Birds will fly high and drop nuts, cracking them open on the pavement. Bears use straws to suck termites and ants out of the insects' habitats. Certain primates know how to separate grain from sand by throwing the gritty mixture into a pool and letting the sand sink while the grain floats. Animal intelligence is revealed also in deceitful behavior. Deceit requires a calculating approach that takes others into account; you can't trick someone (or something) unless you can forecast that someone's (or something's) reactions.[36]

We infer from their behavior that many nonhuman animals also have some sense of themselves as individual entities. They are able to separate themselves from other members of their species, other species, and the environment.[37] A clever demonstration of this was provided by Gordon Gallup, who put small dabs of paint on some chimpanzees' foreheads. The ones that saw themselves in mirrors appeared to be concerned about this and attempted to remove the spots. Chimpanzees can monitor themselves on TV, even when the size of the image is varied or when a left-right reversal (a mirror image) is displayed. However, the same type of testing doesn't work well with dolphins, leading some researchers to conclude that a "simple extension of the tests for primates may be less than ideal for significantly different species," including extraterrestrial species.[38]

From the human perspective, the complex system of symbols and rules that we call language is the hallmark of intelligence. A symbol is anything spoken or written in such a fashion that it is taken to represent something else. Syntactical rules are agreed-upon procedures for combining words to convey meaning. Language frees us from some of the limitations imposed by space and time, since we can discuss events that occur at other places and other times. Because of language we do not have to depend on "try it and see" learning procedures, since other people can tell us what to do. And, of course, language helps us communicate with, and learn about, one another.

Scientists and animal trainers have attempted to teach animals, usually primates, to speak. Many such attempts have been disappointing and tell us more about the psychologists' theories and expectations than the minds of the organisms under study.[39] Early attempts, which involved raising apes as if they were children, led nowhere. Although, on occasion, one animal or another appeared to recognize a word or even uttered sounds reminiscent of such words as "momma" or "poppa," their performances were not compelling and they were rapidly outstripped by humans of the same age. Apes and chimpanzees, however, lack the vocal apparatus to produce human words. A second generation of studies, which took the nonhuman primates' limited vocal abilities into account, met with greater success. Through gesticulating, signing, manipulating objects, using simplified keyboards, and the like, nonhuman primates can, in limited ways, express themselves, make and respond to demands, and describe simple relationships such as "same" and "different." Furthermore, they can use these tools to interact with each other as well as with human experimenters. But what if animals could speak?

Irene Pepperberg gave intensive instruction to a gray parrot by the name of Alex.[40] For over 15 years, Alex has been raised in a rich environment where researchers invite him to participate in their communicative activities. Earlier, unsuccessful attempts to train parrots to speak involved Skinnerian or operant-conditioning procedures in which experimenters used bananas or other rewards to reinforce appropriate verbalizations. Pepperberg's research, which relies heavily on social learning or modeling, has gotten better results. Alex has developed a vocabulary of approximately 85 words that can be strung together into sentences. He understands three categories: size, shape, and color. Alex earns good scores in assigning objects to categories. He is not limited to stating "same" or "different" when given a pair of objects. Instead, he can indicate in which ways these objects are the same and in which ways they are different. For example, a penny and a dime are similar in shape but dissimilar in color. Alex can work with new as well as familiar stimuli, and simple conditioning does not seem to provide a convincing explanation of his proficiency.

Earlier convictions that parrots could mimic vocalizations but not speak rested not only on the failure of operant-conditioning techniques but also on the parrots' lack of a significant cerebral cortex. But it is obvious that some other part of Alex's brain, most likely the striatal regions, serves his speech. This is not to say that animals have the same linguistic skills as humans, but it is to say that they not only communicate with one another but to a limited extent also manipulate symbols and follow rules.

While the perceived intellectual distance between humans and other animals may be diminishing as a result of such findings, humans remain the only terrestrial species capable of mounting a search for extraterrestrial life. It may be, as Dean Falk suggests, that because of competition and struggles for dominance, each planet can host but one species capable of developing an advanced technology.[41] What if humans had not evolved? Paleontological evidence hints that a dinosaur not much larger than ourselves was walking on its hind legs and had developed the equivalent of a semi-opposable thumb and forefinger prior to its extinction (along with all other dinosaurs) about 65 million years ago.[42] Computer modeling suggests that if this *Stenonchyosaurus inequalis* continued to evolve as it had, it would bear a striking physical resemblance to humans.[43] Although there would have been biological limits to their plasticity (at the time they died off they approached the intellectual level of a possum), it is tempting to endow their imaginary descendants with the brain power to achieve our level of technology.

Chris McKay once remarked that, for all we know, the dinosaurs *did* evolve to our intellectual level, or beyond![44] His point is that civilizations disappear without a trace. During the millions of years since the dinosaurs' extinction, continental drift, plate tectonics, wind and water erosion, and the like would have pulverized any artifacts or other signs of their civilization. This could have happened on other planets in our solar system—one theory is that Mars was the site of life at an earlier time—or almost anywhere else in the universe. Evidence of humans is likely to persist for millions of years, however, at the *Apollo* landing sites. Since the moon is inactive, and since there is no atmosphere, the astronauts' footprints and artifacts will be preserved forever.

Dimensions of Intelligence

In the 1920s, Charles Spearman suggested that intelligence was a general ability, a broad capacity of the individual that manifests itself in a wide range of problem-solving situations. From his perspective we can order people along a single continuum of intelligence, ranging from the brain dead to our greatest geniuses. A person's location along this continuum would provide a good indication of how he or she would be likely to perform in any situation requiring intelligence: assembling a motor, giving a speech, making change, composing or humming a tune.[45]

Today the evidence suggests that intelligence is multifaceted and has several components.[46] Consequently, a person or other organism can be highly gifted in one area but not in another. Some people are better with words than with numbers; others are better with numbers. Some people are better at manipulating objects than concepts; other people are better at manipulating concepts. An extreme case is the savant who can do something truly spectacular (such as multiplying two ten-digit numbers in his or her head) while somehow seeming to be deficient in all other areas of intelligent endeavor.

Perhaps we tend to look for certain kinds of intelligence and ignore others. This is suggested by the case of "Clever Hans," a horse that attracted some note in Germany shortly before World War I. Hans seemed able to answer questions about dates, musical harmony, and the location of objects, and he could do simple math problems, including the extraction of square roots. After the questions were phrased in German and written down on a slate, Hans would reply by pointing his nose, shaking his head yes or no, and tapping with his hoof. Astute observers theorized that Hans's trainer signaled the horse when to stop clopping but found that Hans gave other examiners the correct answer even when the trainer was out of the room. Ultimately, it was discovered that Hans could perform well only if certain people were present and if the person who asked Hans the question knew the answer himself. Apparently, Hans was able to tell by nonverbal cues such as changes in posture and subtle gestures where to point his nose, when to shake his head and nose, and when to quit tapping his hoof. Hans did not demonstrate that horses could do simple math, but typically overlooked is Hans's amazing ability to read subtle nonverbal cues.[47]

Different competencies are related to different structures in the brain. Joseph Royce was among the first to stress that whereas there may be superbeings that outperform us on all dimensions, it is perhaps more reasonable to expect that they will outweigh us in only *some* areas.[48] That is, we might encounter creatures that have breathtaking prowess in some areas but appear average or even deficient in other areas where humans do well. Thus, we need to be careful when we use such labels as "intelligent" or "unintelligent." "By being less quick to fix on one-dimensional paradigms that rank pan-stellar brain power in terms of smart and dumb," writes Timothy Ferris, "we will become a little less dumb ourselves."[49]

Human discussions of animal intelligence tend to focus on what we think we do well and then argue about how far animals fall short. But the shoe could be placed on the other foot. One might imagine, a bird with a good compass or map sense—able to navigate by visual landmarks, odors, position of the sun, and the passage of time, or able to respond to magnetic fields—decrying human tendencies to get lost. Perhaps octopuses, which simply sleep until their favorite prey stumble along,[50] wonder about human anxiety, impatience, and willingness to expend energy for second-rate food. Or, perhaps dolphins, which let one hemisphere sleep at a time,[51] would argue that any species such as humans, who are completely detached from reality eight hours a day, could not possibly survive.

Intellectual Development

As we progress from infancy to old age, the ways in which we think change. During our early years we become better able to deal with large amounts of information and solve increasingly complex problems. In general, by adolescence people have achieved the cognitive or problem-solving skills that will serve them until old age. This does not mean that we know everything or will not learn new ways to solve problems or get good results, only that the most basic methods for finding, assembling, and analyzing information are in place. The infant may have difficulty with hand-eye coordination; the adolescent's approach to problem solving is based on the same strategies as scientists use.

There are two basic perspectives on the development of intelligence. One is the incremental, gradualist, or continuity approach, in which skills, abilities, knowledge, and other essentials are thought to accrue slowly over time. From this viewpoint, differences in intelligence are largely matters of degree. In contrast, the stage theory posits periods of rapid change followed by periods of stability: intellectual development is analogous to a person who slowly ascends stairs, stopping briefly on each step. Rapid development is marked by each step up; relative quiescence is marked by the brief pauses.

Probably the best-known stage theorist is the Swiss psychologist Jean Piaget, who identified four stages of intellectual development. Associated with each of these stages are *qualitative* differences in thinking and problem solving. That is, infants, younger children, older children, and adolescents actually think in different ways. The stages themselves are entered in an invariant sequence and are the same across cultures.[52]

John Baird observed that it does not take a great leap of the imagination to conclude that different species, with different genetic plans and maturational rates, undergo a different and perhaps extended series of stages.[53] Piaget believed that the older child can master problems beyond the grasp of the younger child. Might not a superintelligent species have developed ways of thinking that are far beyond our own? Could Piaget's fourth, final, and most exalted stage of mental development be but one minor rung on their ladder to intellectual maturity? We might be like toddlers to them, or the two species may differ so substantially that there is no overlap in the stages whatsoever.

In her book *Human Minds*, Margaret Donaldson presents an interesting developmental model that may be of broader applicability than her title proclaims.[54] This model draws on both thought and emotion. "Thought" refers to information processing, and growth is in the direction of new, more abstract thinking. "Emotions" power or motivate our activities; if something doesn't matter to us one way or the other, we won't take any action.

Drawing on many sources, Donaldson reminds us that for the first few months of life, human consciousness is focused on the here and now, a small pinpoint in time. As we mature, we become capable of remembering the past and anticipating the future. Prior to eight months, a baby will act as though a toy that he or she has seen placed behind a screen doesn't exist; after eight months, the baby will look for the toy and try to retrieve it. The simple act of looking for the toy requires a recollection of the past (remembering that the toy was visible earlier) and the future (an anticipation of playing with the toy, which motivates the search for it). With increasing age, a person's time perspective grows longer; by five or six years or so, children refer to past and future events in a convincing manner.

Over time, there is also a broadening of thought. Initially, the child is highly egocentric; everything is interpreted in terms of its relevance to the self. But as development proceeds, the child becomes aware of other people and places. The ability to think in terms of the "here and now" is supplemented by the ability to think in terms of the "there and then." With continued development the ability to think in terms of concrete instances is supplemented by the ability to think in terms of suppositions (some place, some time) and, ultimately, abstractions that transcend space and time (out of space, out of time).

Through our ability to transcend space and time, we find universal laws and develop broad ethical principles. This broadening of thought is reflected in extending beliefs from "It is wrong for me to attack the person who is frustrating me" (here and now) to "It is wrong for Judy to hit Punch" (there and then) to the general principle that "violence is never justified" (out of space, out of time). Humans did not always have the ability to work in the more advanced modes, but the Babylonian theory of numbers, coupled with sophisticated algebraic manipulations, suggests to Donaldson that the ability to think transcendentally was in place at least 3,500 years ago.

Although we humans have proven ourselves capable of everything from single-point to transcendent functioning, different modes are encouraged by different cultures and subcultures during different historical periods. Thus, in some cultures and at some times, support for transcendent functioning has led to advances in science and ethics: Greece during the Golden Age, the European Renaissance. In other cultures and at other times (in Europe during the Middle Ages, for example) there has been a relatively narrow focus.

No culture, terrestrial or extraterrestrial, could become spacefaring or initiate a radio search if its member's minds were constrained to that narrow pinpoint, the here and now. The ability to draw on the past and to anticipate the future, to think transcendentally, is essential for its science. And, in the same way that Donaldson's framework encourages us to make certain basic inferences about the intellectual powers of aliens, it encourages us to infer that they will have emotions. In the absence of feeling states they will not be motivated to act in ways that will reveal their presence to us. They must be motivated to build broadcasting stations or interstellar arks. We may surmise that even the cold and emotionally detached alien abductors must have feelings inside; as they perform their cruel gynecological exams they must feel good, or at least better than they would have felt if they had left their alleged victims alone.

CONCLUSION

Earth supports an incredible array of organisms that differ along many key dimensions, including size, shape, form, physical strength, intellect, and temperament. At the same time, these diverse life-forms have certain similarities, and it is what they have in common that gives us a springboard for generating hypotheses about extraterrestrial life. The requirements for life itself, coupled with the biological and psychological requirements for conducting an interstellar search, impose certain constraints on the kinds of beings we are likely to find in the course of a microwave search. Because of this, SETI, the search for extraterrestrial intelligence, might be known better as SETILO, or the search for extraterrestrial intelligence like ourselves.[55]

At the most fundamental level, extraterrestrial life-forms can be analyzed in terms of size, shape, and functional subsystems that process some combination of matter, energy, and information. The requirement for technology eliminates organisms that are too small or weak to tend fires and work metals. The requirement for sufficient intelligence and technology to construct a powerful microwave beacon eliminates tiny, undifferentiated organisms. The upper limit is less clear, but disproportionately large skeletons, tremendous energy requirements, and other factors may make it difficult for truly immense creatures to signal across interstellar distances. Additionally, we should not expect to hear from organisms that lack intelligence or language, that are rooted in one place, or that are continually on the move to satisfy the basic requirements of existence. Because we expect to encounter a civilization much older than our own, we are concerned that our intellectual differences will militate against communication and understanding.

In a series of thoughtful articles, E. J. Coffey advances the thesis that, since chance plays a major role in evolution, we cannot make good predictions of the putative extraterrestrials' nature.[56] Coffey points out that different ecologies give rise to different morphologies and different behaviors. Since terrestrial and extraterrestrial life-forms will not have common ancestors, and since we cannot begin to imagine their environmental realities (never mind the chance events that occur within them), we cannot make reasonable guesses as to what they will be like. Unlike evolutionary biologists, who must stick to the data, physicists and other SETI buffs engage in a dangerous game of "attributing *hypothetical* behaviors to *hypothetical* creatures with *hypothetical* bodily characteristics."[57] In essence, argues Coffey, SETI enthusiasts design a make-believe creature that suits their purposes and then assume that evolution brought it about. He argues that it is unlikely that evolution elsewhere will produce a humanoid alien. And since function follows form, we should not delude ourselves that a nonhumanoid life-form will show humanlike behaviors. SETI scientists' attempts at "alien design," he says, bear no relationship to the data and represent instead the psychological process of projection, whereby we see in others (i.e., the aliens) a reflection of ourselves.

While a living-systems analysis suggests many similarities between terrestrial and extraterrestrial life-forms, it does not deny diversity in the universe, nor does it imply that "they" will look like or be put together in the same way as "us." We do not have to postulate that humans and aliens have common ancestors or that they undergo a common sequence of evolutionary events to generate plausible hypotheses. It is not necessary to become embroiled in whether they will look like apes or reptiles or giant slugs. Instead, we should seek analogous structures and processes. If they are not composed of atoms and molecules; if they lack the ability to reproduce or move; if they cannot ingest, store, and use matter, energy, and information; if they cannot adapt to environmental change and keep their internal states within certain limits—then and only then will they will fall beyond Living Systems Theory's enormous reach.

SIX

SOCIETIES

Many people think of extraterrestrial societies as almost utopian. Abundant resources, efficient production, and material wealth come to mind. High technology is a given, since any society that we will contact through a SETI-like search must have a level of technological sophistication that is at least equal to our own. Such societies will not only enjoy high technology but will have solved all their social ills as well.

Just how advanced are they likely to be? Hardly anyone expects them to be less than 10,000 years further along the path than we, and there are guesses of up to 1 billion years. These figures are based on estimates about when extraterrestrial societies are likely to have emerged and how fast they developed technologically. Once societies are formed and work their way through technological adolescence (atomic weapons, pollution, overpopulation, genetic deterioration, and so forth), they may persist indefinitely, with the result that many ancient societies may continue to exist today.

Some of our reverence for these truly hardy societies may rest on an assumption that technology will increase linearly, or perhaps even exponentially, over time. Overall, human progress in science and technology has been uneven. The Golden Age of Greece and the Renaissance were punctuated by the Middle Ages. With the exception of Egypt, science, as many of us think of it, barely got off to a start in Africa. Science and engineering got off to good starts in China and India, but stalled hundreds of years ago. During the 1960s the U.S. government couldn't throw enough money at research; during the 1990s it is difficult for even talented and persistent researchers to pry money loose.

Inventions and discoveries are sometimes lost, and they are not always found again. Although we believe that the world's ancient monuments such as the pyramids and Stonehenge were built by the world's ancient citizens, we aren't completely sure how they did it. Skills that were available at the dawn of the 20th century are disappearing—it may now be impossible to replicate the Chrysler or Empire State buildings or other skyscrapers of the 1930s and 1940s because too many critical crafts such as masonry have gone to the grave with their practitioners. (Thus future generations may have to settle for steel girders hung with glass.) Compared with 50 years ago, it is more difficult now to build a battleship like the *Missouri* (among other things, there is no way to cast the huge gun turrets), and we would have to retrace many steps if we ever choose to build another Saturn V rocket. Were the "Dark Ages" an aberrant depression on a steep path to technological sophistication, or is the recent flurry of progress an abnormal "spike" on a meandering path?

At least three factors can retard the march of science, technology, and discovery. The first is the society's material resource base. Some advances may be judged too expensive, at least relative to expected gains. The second is social policies that neglect science and invention, discourage inquisitive behavior, or decree that resources should be used in other ways. Kenyon B. DeGreene suggests that within organizations, dominant coalitions "select and implement those technologies that do not threaten its own integrity, that reinforce the existing enterprise culture and practices, and that reduce variety and freedom."[1] Finally, the most advanced society's progress must be capped by the physical nature of the universe itself. Even with virtually unlimited resources and time it might not be possible to achieve faster-than-light travel. We may be dazzled by extraterrestrial technology, but the point remains that even if their civilization is a million years older than ours it is not necessarily a million years technologically more advanced.

Continued societal evolution could lead to moral and social as well as technological development. The freedom from want resulting from an immense resource base and high technology, coupled with the ability to think transcendentally, could lead to high ethical principles and enlightened social policies (see chapter 11). As J. W. Deardorff points out, the assumption of mediocrity, which suggests that humans have only an ordinary intelligence, may apply to human values (ethics, social responsibility) as well.[2]

Shirley Ann Varughese proposes that we grade societies on two dimensions, technological and social.[3] By her definition of development, societies that have achieved only a low level of social development are likely to be unstable, conflict-laden, and warlike. Those that have achieved a high level of social development are likely to be stable and peaceful. The most dangerous society would have high technological but low social development—that is, great ability to inflict harm but little self-control. However, as we will see later, irresponsible societies may put themselves out of business, with the result that we are unlikely to hear from societies consisting of morally bereft technological geniuses.

At the level of the society, reproduction is accomplished by subsystems that set up new societies modeled on the old: a revolutionary council establishes a new charter, military conquerors "reform" a subjugated enemy, and so forth. Examples of new societies might include orbiting colonies or lunar or planetary bases elsewhere within the local solar system or anywhere else that transportation systems permit. Ultimately, small, football-size probes containing DNA codes could distribute a species throughout the galaxy or universe.[4]

CULTURE

Culture plays a major role in many discussions of extraterrestrials and their societies.[5] *Culture* is an organizing concept that anthropologists use to refer to the human characteristics of a group, organization, or society. As John Berry and his associates note, "culture" has been defined, variously, as "a shared way of life," "that complex whole which includes knowledge, beliefs, art, morals, law, customs and any other capabilities and habits acquired by man as a member of society," the "total social heredity of mankind," and the "man-made part of the human environment."[6] Culture is an emergent, or superorganic, concept; sometimes it is viewed as the "personality" of a society. Culture has a long past and imparts stability to the future. As members move away or die, the culture perseveres because there are new generations to carry on traditions. Culture arises from a mix of people, tools, and symbols, just as life might arise from an appropriate mix of chemicals.

Paul Bohannon reminds us that culture is an integral part of nature and that cultural variables are continuous with physical and biological variables. Matter is a basis for life, and life provides the foundations for culture. Culture transcends and enriches matter and life, but it does not change the way that physics, chemistry, or biology work. We must resist the temptation to see culture as an artifice, somehow discontinuous with or separate from nature.[7]

Cultural systems contain the products of action as well as the processes that condition future action. Brain growth and intelligence make it possible for culture to serve this adaptive purpose for humans.[8] Culture represents a form of adaptation to ecological and sociopolitical factors.[9] Cultural transmission makes it possible to pass on detailed information about a complex, rapidly changing environment. It allows our species to develop much more rapidly than it would if we were forced to rely on behavioral changes brought about through biological evolution.[10]

Culture is a pervasive and powerful determinant of behavior. Its hold over us ranges from the highly prescriptive to the merely suggestive, from immutable principles that must be obeyed at all cost to general guidelines hinting that one option is preferable to another. People within a culture are exposed to it from the earliest stages of life, and both formal and informal mechanisms ensure that it moves from generation to generation. Because the influence begins early and continues for life, it becomes highly ingrained, and culturally conditioned behaviors are almost automatic. Only occasionally do we have to riffle through a rulebook or seek advice before deciding what to do.

Sandra and Daryl Bem once referred to our beliefs about the appropriate (that is, sex-typed) roles for men and women as "nonconscious" in the sense that we never really think about them.[11] Because of this, the reactions we have to men and women are rarely examined, and they are perpetuated across generations. Anthropologists such as Bohannon believe that many of the mores and folkways that so deeply affect human lives are similarly nonconscious or automatic. That is, the assumptions that underlie our value judgments, proclivities, and activities are rarely examined, and much of what we learn is applied in a mindless way. When we are operating in our own culture, we do not have to think twice about which hand to extend for a handshake, about which side of the road to drive on, or whether to use utensils when consuming a particular type of food.

In effect, culture is transparent: it is invisible to the people who are within it. The risk is that we may mistakenly apply our cultural framework to ETs beliefs, attitudes, or behaviors, or take for granted that ETs will somehow understand the framework that guides our own thoughts, feelings, and actions.

Universals

Anthropologists have devised taxonomies to organize volume upon volume of descriptions of Earth's many cultures. Some of these frameworks may be fruitfully applied to nonhuman species. Particularly promising are the anthropological studies that have led to the identification of "cultural universals," that is, characteristics that appear with such regularity as to suggest that they are essential for all terrestrial societies. Some of these are of broad applicability; for example, Bohannon identifies four "functional prerequisites" that apply to both human and animal social behavior: dominance, kinship, the specialization of tasks, and cooperation.[12] Because these cut across both cultures and species they satisfy our search for universal principles.

All animals assert themselves to obtain living space and the resources necessary to sustain life. Cross-species studies suggest certain benefits associated with an unambiguous ranking system. Although one might expect that societies with large power differentials and sharp status distinctions would be more riddled with disputes than those where the pecking order is unclear, the reverse is the case. There may be initial fussing and fighting as a hierarchy is established, but once it is in place everyone understands who can lick whom, and the dominant members of the group are only infrequently called upon to assert their authority. When they are challenged, they may resolve the difficulty with threatening gestures and displays; they have little need to rely on actual force. In less hierarchical species where dominance is not established, there are pressures to "have a go at it" to resolve even small issues.

Societies, as well as individuals, sometimes resolve issues through puffery and threat. Wars do not have to be fought to the death or to unconditional victory. In some clashes, two armies (or groups of warriors) gather and, with much energy and noise, begin to fight, but the ruckus ends as soon as one warrior is killed or even seriously wounded. Then, the two sets of combatants may fraternize and perhaps even enjoy a feast.[13] One can imagine the consternation of warriors from a culture where combat is ritualized who find themselves fighting opponents intent on their annihilation.

In human societies, individuals defer not only to a social superior but also to social institutions and the community or society as a whole. Social psychological studies of conformity show that people will frequently "go along with the crowd." In the 1960s, social psychologist Stanley Milgram conducted a series of experiments in which perfectly normal individuals were led to believe that they were inflicting painful shocks on someone else. There were several variations of these experiments, but the basic finding remained: people obeyed the experimenter's orders to deliver harmful, even potentially lethal, shocks to another human being.[14] This was only an experiment, and despite impressive equipment and sound effects, no shocks were actually delivered. But in other cases obedience has led to massacres and even genocide. This is reflected in the attitudes of some Nazi war criminals who explained themselves by saying they were "only following orders."

Although society may require subordination of the individual will, there must also be room for individual initiative and action. Independence is required for innovation, and people must have a little slack to satisfy their personal needs. One goal of societies, then, is to impose sufficient constraints so that individuals can live and work together, but at the same time leave latitude for individuality, creativity, and freedom of expression.

Discussing his experiments, Milgram suggested that social organisms must be capable of functioning in two states. The first is the individual state, where people believe they make deliberate choices and then follow through. The second, or "agentic," state involves functioning as a part of the larger social organism. The key to entering the agentic state is one's perception of another person as having the right to demand compliance. This may occur when the person who makes the demand is in a position of authority by virtue of research credentials, or it may be triggered by a symbol such as a uniform or badge. Extraterrestrial societies, like human societies, will have to solve the vexing problem of establishing a social order that is sufficiently rigid to ensure coordination but flexible enough to accommodate innovation and change. Perhaps, like many terrestrial societies, their society will be a mosaic of subcultures. Evolution selects for entities possessing divergent traits and behaviors, and for this reason, over the long run monocultures may be at a disadvantage.

A second universal is some sort of kinship system. In order to survive, members of all animal species must engage in sexual reproduction and parent their offspring until they are ready to strike out on their own. Among humans, kinship systems include biological kin plus other people who have opted to treat one another as kin. Sexual activity and favoritism perpetuate one's genes through extending and bettering the family.

Our knowledge of different species on Earth may yield hypotheses about extraterrestrial reproductive and parenting practices. Here, species that have relatively small brains and low intelligence tend to have short life spans, brief gestation periods, and large litters of offspring. Their young mature rapidly and require little parental care. Much of their behavior is prewired and preselected. Species such as primates and cetaceans that have relatively big brains and high intelligence tend to be relatively large and have long life spans. For the most part, they have single offspring or very small litters. The young mature slowly and require extended parental care. Their large brains require significant "programming," whether we construe this as analogous to inscribing knowledge in the brain or as selecting from a large number of sophisticated prewired options. The general rule is that it takes time, care, and attention to produce intelligent offspring.[15]

Kinship may help us understand certain acts of self-sacrifice if we think of evolution as a game where the team goal is to pass the maximum number of genes on to the next generation. Biological or social factors may prevent a specific individual from reproducing, but he or she can take actions that increase the chances of a blood relative's survival. We know, for example, that animals will take risks to protect progeny that they would not take to protect unrelated animals. Furthermore, humans will give forms of assistance to blood relatives that they will not offer to nonrelatives. The bachelor who leaves his fortune to his nephews and nieces will help perpetuate the family line even if he has no children of his own. Thus, acts of individual sacrifice may increase the likelihood that some of one's genes will pass on to subsequent generations.[16]

Responsibility to our kin may extend, if in an occasional and imperfect manner, to the larger community or even to people in general. Of course our society has mass murderers, child abusers, and other irresponsible people, but on the whole people within a society are expected to protect one another, even if this involves taking aggressive action against an outsider. Our level of interest in responsibility is underscored by the huge resources that we are willing to devote to counteracting irresponsibility by funding mental hospitals, police agencies, prison systems, and rehabilitation programs.

Useful here is Erik Erikson's idea that, in all cultures, children around the age of seven or eight develop a "sense of industry," that is, a desire to do something worthwhile and make a positive contribution, although the precise nature of such activities may vary from culture to culture. (In our culture a sense of industry includes continuing school, getting a job, and advancing up the occupational ladder.) This does not mean that each culture is fraught with Type A personalities and overachievers, only that there must be enough of a sense of industry to perform the society's basic functions.[17]

Specialization is another prerequisite for culture. Human societies (and many other "societies," including those of ants and bees) are characterized by a division of labor such that individuals specialize in only one task or a very limited range of tasks. The prototype is found in the different sets of activities required for two animals to mate and raise offspring. The degree of specialization depends on the size, complexity, and technology of the society. In contemporary human society, specialization, or "the division of labor," is most obvious in the sphere of work. No individual can begin to provide the many goods and services required by an advanced society, yet all of these can be produced by the aggregate. In addition to making large tasks possible, specialization helps people work efficiently through a repetition of familiar, well-learned tasks. Moreover, specialists can learn their jobs in great depth. Specialists play particular roles in society, and it is in large part the aggregation of the specialists' activities that determines a society's fundamental nature.

At least some of the time, members of a given culture must cooperate, that is, coordinate their activities for mutual gain. Cooperation is the foundation of many forms of social life. One of these is mutual defense against rival groups of the same species (other baboon troops, other street gangs) and against predators. Another is the rearing of young. Communal efforts are efficient because a relatively small number of adults can care for a relatively large number of infants. Finally, collaborative activity multiplies and diversifies the results of individual effort. Teamwork makes it possible for predators to capture prey that are too large or have too many defenses to succumb to an individual. As John Alcock points out, it is not uncommon for predators to capture prey that is 6 to 12 times the predator's size.[18] This not only results in more total food for the group but in more food per predator.

On Earth, and perhaps everywhere else, combining individual efforts is a prerequisite for advanced technology. Some people have greater insights and carry their ideas further than others do, but on the whole, technological advancement involves collaboration and the gradual accumulation of results. Even the Leonardos, Edisons, Fords, and Whitneys of this world drew on the ideas of their predecessors and contemporaries. Today, cutting-edge science is big business, and it is not uncommon for scores of high-energy physicists to collaborate on one project.

Thus, social living sets the stage for cooperation as well as for competition or aggression. In some cases, one organism may exploit other organisms by not doing its share of the work while nonetheless receiving full benefits. In human societies, there are many rules, not always effective, intended to limit competition and prevent outrageous exploitation. Groups and communities often bring strong pressures to bear on those whose behavior is not up to code.

ET Cultures

We have good reason to believe that when we intercept a transmission we will have encountered another social species. A single organism is likely to be wholly occupied by activities to sustain its life. A large group can divide activities with sufficient efficiency to leave enough "human resources" (or at least living resources) to pursue scientific inquiry. Reclusive species are unlikely to want to know about creatures from solar systems such as ours and would be unlikely to initiate a search. When we add in the notions that science is a collective activity and that only a large, complex society can afford the luxury of a search, the best guess is that ET's microwave beacon will be a group effort backed by the accumulated knowledge and resources of a society. The extraterrestrials will be social, they will have tools, they will have language, and because these and other elements are present, they will have culture.

Furthermore, when we encounter their culture, we will have some basis for exploring it. We should begin by looking for features that are universal on Earth and by applying categories that have proved helpful in wide-scale comparisons of human cultures. Of course, there is always the danger that preexisting conceptual schemes will blind us to other considerations that are crucial for understanding their way of life. However, such prerequisites as dominance, kinship, specialization, and cooperation are sufficiently broad-based as to provide useful starting points.

MATTER–ENERGY PROCESSING

According to Living Systems Theory, societies, like organisms, process matter, energy, and information. Societal subsystems that process matter–energy include industries, firms, and agencies that obtain, store, process, and distribute products and raw materials. Usually these subsystems serve the society's central productive function, the creation of material wealth.

As Adrian Berry points out, technology and economic prosperity advance hand in hand.[19] Both general wealth and the quality of the average person's life have increased tremendously over the past few centuries. A highly productive, efficient food industry, coupled with an excellent distribution system, provides us with an unending variety of good food, in or out of season. Upon seeing the quantity, quality, and freshness of our food, it is doubtful that a soldier from a century ago, accustomed to hard tack and salt pork, would share our concerns about whether it was organically grown or laden with excess salt, sugar, or fat. The work week has shrunk, work is less dangerous, many of us have paid vacations, and many of us can look forward to a long and prosperous retirement. Everywhere we look—transportation, education, housing, health care, recreation—advancing technology has made things better and better. Despite conjectures about dying planets and the like, our experience thus far suggests that advanced technological societies are likely to be characterized by high prosperity and a high quality of life.

Resource Bases

Terrestrial societies occupy land mass, but a more advanced society may be situated on a large orbiting satellite, a self-contained "worldship," or another construction. The boundaries of terrestrial societies are usually marked by geographic features, such as bodies of water, mountain ranges, or great walls, and they are always defined politically. A society's boundaries are enforced by organizations such as immigration services, border patrols, and the military. Like all other open systems, a society has semipermeable boundaries; that is, boundaries loose enough for the importation and exportation of matter, energy, and information, but tight enough to separate it from neighboring societies.

Although a society may cover a certain area of land, boundaries can extend outward into territorial waters and upward into territorial airspace. When we look up, we might imagine two boundaries of differing degrees of permeability. Each of these can be conceptualized as gradients of different densities, the matter boundary dense and compact, the information boundary thin and dispersed. The matter boundary is fairly impermeable due to the requirements of lifting matter into orbit and transporting it over interplanetary distances. This boundary forms, for all practical purposes, a relatively tight sphere around the home planet, perhaps allowing the transport of matter and the beaming of commercially useful amounts of energy to the limits of that planet's solar system. The information boundary is relatively permeable, since information can be transmitted at the speed of light and for relatively modest investments in energy. This information boundary forms a very large and amorphous sphere, permitting, perhaps, an intergalactic exchange of information, depending on the availability of power and the sensitivity of the receiving apparatus.

Several eminent space scientists, including Konstantin Tsiolkovsky and Krafft A. Ehricke, have a strong philosophical and humanistic as well as scientific bent and argue that humans are destined to migrate into space. Tsiolkovsky's credo, "The planet is the cradle of intelligence, but it is impossible to live forever in that cradle," reflected his belief that movement into space was essential for further human evolution.[20] Ehricke formulated an "extraterrestrial imperative" that urged human expansion into space.[21]

Ehricke believed that migration to the stars will stretch people's physical, mental, and spiritual capabilities. Moving into space through a logical and graduated series of steps creates not only a mind-set but also economic and cultural conditions that foster personal and societal well-being and encourage a self-perpetuating, positive spiral.

As a growing population on Earth taxes the planet's resources, we are confronted with two choices. One is a policy of moderating our growth, of limiting human reproduction and rationing Earth's resources so that they will last a little longer. This choice, believed Ehricke, will reduce the quality of life of the average person and promote negative attitudes toward our main hope for salvation, science and technology. The preferred choice is to expand into space, thereby increasing the resource base so that we can continue to consume, enjoy, and expand. Migration into space, Ehricke argued, encourages a healthy "can do" mentality, an international perspective, and the continuing development of science and technology, which in turn provide the key to an ever expanding array of opportunities and successes. "Thinking small" treats the Earth as a lifeboat, as a closed system characterized by chauvinism, repression, stagnation, decay, and an inexorable slide toward the death of our biosphere. "Thinking big," according to Ehricke, expands our resources, eliminates the causes of war and revolution, and leads to continuing economic, psychological, and social development.

Because we ourselves are poised on the verge of space industrialization, we can easily imagine the development of an extraterrestrial civilization: population pressures, coupled with an increasing ability to exploit the power of the sun and other space resources, would encourage migration into space. Extraterrestrials could have migrated to orbiting colonies, lunar bodies, and other planets. They could mine the asteroids and use space platforms to harness their sun's energy. From the perspective of a society that is further along than ours, human attempts to move into space must appear very feeble. Artificial means are required to meet life's minimal requirements. Danger, limited supplies, a lack of amenities, and close confinement yield a low quality of life for today's human spacefarers.

Space settlement offers a species new possibilities. Discussions of future human space settlements tend toward the ideological and visionary.[22] Usually, this ideology calls for healthy living, a very high environmental quality, and the replication of both Earth's archetypal settings and endearing natural phenomena (such as hills, clouds, sunrises, and sunsets).

Perhaps the biggest barrier to widespread human movement into space is the sheer cost of lifting materials out of Earth's "gravity well" and into orbit. To counter this we must exploit indigenous construction materials (primarily metals and ceramics) that are already in space. For us this may initially consist of using lunar or Martian regolith (rocky soil) as radiation shields for structures imported from Earth. Later, there will be more extensive use of regolith for surface structures and orbiting satellites. For example, buckets of materials mined on the moon may be individually hurled into space, where they are gathered and used to assemble an orbiting colony. Because of the relatively low gravity on the moon, this can be done by an electromagnetic catapult known as a mass driver.[23] Later will come mining, smelting, and manufacturing operations to extract construction materials from asteroids and planets.

When we are able to do this, whole towns and even cities may be developed from scratch, and it will be possible to follow thoughtful, planned, holistic approaches instead of the haphazard approaches that so often predominate on Earth. When we can choose among planets and asteroids (and customize those that are unsatisfactory), we will no longer be constrained by such "givens" as external temperature, humidity, atmosphere, gravitational pull, length of day or year, or much of anything else. Whereas today's architects must work within greater constraints in space than on Earth, tomorrow's architects may find that Earth imposes more limits on their creativity.

Since the earliest of times community and habitat designs have both reflected and shaped the interpersonal and sociopolitical behavior that occurs within them. Magorah Maruyama believes that spacefaring civilizations can choose or even invent cultures, social patterns, and social philosophies, and then develop appropriate material conditions and community designs. For example, the hierarchical and homogenistic community follows a belief in the "one best way" and applies strong pressures to conform. This philosophy might be supported by highly standardized living units. In the individualistic and isolationistic community, independence and self-sufficiency are major virtues. Each living unit is entirely self-contained and, within each unit, everything adjusted to individual taste. Finally, in the heterogenistic, mutualistic, and symbiotic community, variety is a source of strength and enrichment. This type of community might be supported by dwellings constructed on different design principles drawn from different cultures.[24]

Futuristic space stations, lunar bases, and other space habitats may be sophisticated environments that monitor and support their occupants.[25] Sensors may make it possible to track the flow of people, information, energy, and matter. In automatic mode, the system will be self-diagnosing and self-correcting. An emerging theme is that rapid advances in artificial intelligence, automation, and robotics have blurred the boundaries between humans, automated systems, and environments, and have led to the emergence of complex overall systems with higher-order properties. Perhaps such systems should be viewed as *social cyborgs*, that is, advanced fusions of society and technology. The design challenges are first, to develop habitats capable of monitoring and (through automated systems) responding in ways that support their occupants, and second, to achieve better integration of human and technological subsystems.[26]

At least one design study suggests that an advanced society's transportation system could include saucerlike vehicles. After reviewing reports, photographs, and other evidence pertaining to UFOs, John Ackerman, an aerospace engineer with a major engineering and manufacturing firm, has evolved a design for a saucerlike spacecraft capable of UFO-like performance: levitation, stabilization, high speeds, near-instant turns, and the like. Furthermore, he shows how reported changes over time in UFO shape and performance may reflect refinements in this technology. His technology takes advantage of magnetic fields, which pervade all space. The ring around the bulge in the saucer contains a large superconducting toroid that carries an enormous current about its major circumference. This generates a strong magnetic dipole field, which interacts with the ambient fields to produce lift and propulsion. The technologies that would be required, fusion and very high superconductivity, are beyond our current capabilities but not beyond our understanding.[27]

Energy Processing

SETI scientists expect that an advanced extraterrestrial society will make massive use of cheap, abundant energy. First, it will have successfully addressed the scientific and engineering problems that must be solved for safe, reliable, and cheap nuclear power, perhaps based on that elusive process of cold fusion. Second, since their society is likely to be spacefaring, they will have established orbiting power stations that will beam energy from their sun to their planet.

N. S. Karadashev proposed classifying technologically advanced civilizations into one of three types depending on the amount of energy they use.[28] The Type I civilization is at a technological level close to that of Earth and consumes energy at the rate of approximately 4×10^{19} ergs per second. The Type II civilization is capable of harnessing the energy radiated by its own star and consumes energy at the rate of 4×10^{33} ergs per second. Finally, the Type III civilization harnesses the energy of its galaxy, consuming approximately 4×10^{44} ergs per second. By definition, Type II civilizations can restructure planets and Type III civilizations can restructure galaxies. Carl Sagan suggests that the signals of Type I civilizations probably lack the power to reach us, even if (on a cosmic scale) they are relatively nearby. A Type II civilization can communicate across our galaxy and a Type III civilization from across the universe. We may have a higher probability of success by looking for these more powerful societies, even though Type I civilizations may be closer by.[29]

Freeman Dyson has suggested methods for creating a Type II civilization. These involve constructing a huge spherical shell around a sun, so that all of its energy is trapped. The shell is constructed out of readily available materials and need be only 2 or 3 meters thick. Because the light from the encapsulated sun is trapped within the sphere, the Type II civilization would not be optically visible to us, but the shell would give off immense amounts of infrared radiation. Dyson suggests that we seek stars whose motion shows that they have an invisible companion; the companion may be invisible because it is surrounded by a Dyson sphere.[30]

Joseph Baugher suggests that Type III civilizations can control the process of star creation and alter the structure of entire star clusters; can make large-scale interstellar voyages at near-light velocities; and will last forever because when one star dies they either create a new one or move somewhere else. Linked together, the members of such a civilization (which may be cyborgs or even superintelligent computers) form a vast galactic network, thereby creating a superbeing. To us, their works would seem like magic.[31]

The belief that extraterrestrials will have cheap, abundant energy is convenient for those of us who look forward to contact. First, if energy is precious, it is doubtful that they would use it on such high-risk scientific projects as activating powerful interstellar beacons. If energy is plentiful, however, it's no big deal to use some for powerful interstellar broadcasts. Second, even if they are not broadcasting, we might be able to detect their (tremendous) use of energy; for example, it might be possible to identify the signature of stars surrounded by Dyson spheres. A recent search for such spheres in the vicinity of 180 solar-type stars yielded no suitable candidates.[32]

Donald Tarter challenges the view that advanced civilizations will be characterized by immense energy consumption. He suggests that our expectations of Dyson spheres, Type III civilizations, interstellar arks, and the like represent an adolescent view of technology. There may be natural limits to a society's growth, says Tarter, and whereas gargantuan projects such as Type III civilizations may be technically possible, they also have a number of downsides and are in any event prohibitively expensive. Tarter proposes a second model, one based on miniaturization, down to the level of nanotechnology and quantum engineering. This type of civilization will not make itself known through interstellar beacons, or ferocious energy consumption, but through miniature probes and forms of communication that will not be obvious to us. To increase our chances of a successful search, then, we must supplement search strategies based on thinking big with search strategies based on thinking small.[33]

INFORMATION PROCESSING

A society's information-processing subsystems consist of institutions, industries, and agencies that acquire, manipulate, and distribute information. These organizations are analogous to the brain and nervous system of the organism. For example, astronomers, espionage agencies, and international news services bring information into the society; archives and libraries store it; universities, R&D agencies, and governmental decision-making bodies process it; publishers, broadcasters, and telephone networks distribute it. At the level of the society, undercover police, marketing research firms, and public opinion pollsters are the functional equivalent of the organism's proprioceptors.

Advanced Information Technology

Of course many of us expect that, by our standards, extraterrestrial information-processing systems will be highly advanced, in terms of both the complexity and volume of information processed. About one century since the advent of radio on Earth and (using World War II as a benchmark) a little more than half a century since the advent of the first working computers, the real cost has plummeted (as those of us who remember the first home television sets and computers can attest), while reliability, capacity, and versatility have skyrocketed. We can project within the next few years devices that will fit in a car trunk and yet hold an amount of information equivalent to that stored in the Library of Congress.

David Freedman has speculated on future information-processing technology.[34] First, we are close to the limit of miniaturization, at least using conventional techniques. Already individual transistors are approaching the size of single molecules and will have to operate on quantum principles. Second, some of tomorrow's information-processing devices might be characterized more accurately as "grown" rather than manufactured. That is, some information processing will take place in organic cells specifically cultivated for that purpose and some information-storage devices will involve biochemical processes. Third, tomorrow's advanced information-processing systems may be self-organizing.

Work in artificial intelligence (AI) began with a "top down" approach to information processing, which begins at the highest, most centralized, or most abstract levels and then proceeds downward. (In other words, it starts with a few general principles that in combination try to account for the individual case.) Although there were early signs of success, this approach stalled. But the AI movement has been infused with new life by an alternative, "bottom-up" approach, which begins with simple, almost reflexive actions that are then integrated into successively higher levels of abstraction. The bottom-up, or natural, approach can be likened to the evolution of intelligence in living beings and shows high promise for AI applications.

Additionally, notes Freedman, computer programmers have had some success with an evolutionary approach to developing new programs. Programmers begin with many randomly generated programs that ultimately will do a particular job. Those programs are then randomly paired; feedback leads to the survival of some pairings and the elimination of others. Over successive generations, programs become increasingly capable. It is only a matter of time, maybe not too much time, before artificial intelligence can accomplish all the tasks now performed by humans.

Intriguing possibilities arise if, as Tipler[35] suggests, in the very distant future some people may "live" as perfect emulations of themselves in a computer that is giant in terms of capacity if not in terms of size. Specifically, by joining with other emulated organisms, individuals may be able to form a "group mind" with an information-storage and -processing capacity that far exceeds that of any individual. Ultimately, there would be no effective limit to the number of minds (individual computer programs emulating individual persons) that join together. The practical result would be to expand the consciousness of the individuals and create a superordinate entity that will have an ability to apprehend and solve scientific problems of unprecedented magnitude. For the social psychologists of the future, the reality of a "group mind" may be more of an engineering than philosophical issue!

Decision-Making

At the level of the society, decisions are made by individual leaders and by collective decision-making entities such as parliaments, diets, councils, and congresses. Political decision-makers are not entirely free agents in that their decisions are shaped by numerous contextual factors. These include the expectations and interests of neighboring societies (see chapter 7) as well as expectations and interests of special-interest groups and the general populace.

John Baird points out that the probability of our finding extraterrestrials depends not only on their technological sophistication but also on their values.[36] Extraterrestrial societies that broadcast to civilizations such as ours must have the scientific or technological know-how and must also place a premium on exploring space. ETs that are interested in space but lack science may be doing their equivalent of watching for UFOs, engaging psychics, or undertaking other ineffective searches. ETs that are not interested in space but have developed strong scientific techniques will be applying their skills to other topics. Baird's point is that equating estimates of the number of technologically advanced civilizations with the number of civilizations that we might encounter yields inflated results, since some of these technologically advanced civilizations will not be interested in broadcasting or searching.

An extraterrestrial society's huge material resource base, advanced technology, and large (and most likely dispersed) populations may steer them toward certain kinds of political systems. First, as noted earlier in this chapter, dominance is a functional prerequisite for society and culture, and I expect that advanced extraterrestrial societies will have some sort of a hierarchy. At the same time, certain factors may have a leveling effect, that is, they may encourage equality. As already suggested, superior production and manufacturing technologies would contribute to a world of plenty. Advanced agriculture and manufacturing techniques, coupled with a large resource base, should work against poverty and put a floor beneath the lower classes. Superior information-processing systems are likely to contribute to an educated and informed population. These factors would reduce the distance between the ruling and the ruled and, by limiting the power differential, make preserving a monarchy or autocracy an uphill battle.

ET's top leaders may be forced to delegate, and this too could work against a strong autocracy. Centralized decision-making occurs when authority is retained by an individual or a group at a capitol or other centralized location.[37] Decentralized decision-making occurs when the authority to make decisions rests in the hands of subordinates at remote locations or sites, such as governors of distant provinces or starship commanders in outer space. Since such societies are likely to be large and dispersed, they will not do well under the micromanagement of a centralized authority.

Centralized decision-making tends to be effective when the central authority has access to resources that are unavailable in other locations, such as staff experts, sophisticated computers, and large data banks. Effective centralized decisions rest on good communications with outlying regions and with a thorough knowledge of the conditions that are to be found there. Decentralized decision-making tends to be best when the centralized authority has less information than do local commanders about conditions in the field, and when communication is slow or inaccurate. In the latter case the local commander has a better chance of reaching an informed decision and implementing it in a timely manner. Large and geographically dispersed societies are difficult to administer in a centralized manner. Although an extraterrestrial society may possess excellent communications systems, these systems will be limited by real time. Furthermore, leaders at even remote outposts will have access to superb computers and data banks. Consequently, the advantage may go to decentralized decision-making, which in turn reduces the power of the central authorities, another reality that could work against autocracies.

Are there any discernible trends toward a particular political form on our own planet? If we are to extrapolate from terrestrial experience, we must avoid a narrow time span and not be distracted by minor perturbations in the flow of history. We must avoid the temptation to become embroiled in examples and counterexamples and focus instead on broad themes and statistical trends.

Bruce Russett's recent analyses of terrestrial governments reveal an unmistakable trend.[38] This is a shift away from authoritarian forms of governments, and toward democracies. A democracy, by Russett's definition, provides a voting franchise for a substantial proportion, if not all, of its citizens and holds contested elections for an executive that is responsible to the people. Typically, but not invariably, this form of government permits free speech.

The number of democracies in the world depends on one's criteria. By Russett's criteria, only 12 to 15 nation-states qualified as democracies at the end of the 19th century. However, the 20th century has seen a sharp rise: by 1992, almost half of the world's states—91 out of 183—met Russett's criteria. An additional 35 nations were in transition from autocratic to democratic governments. Thus, by the early 1990s, two-thirds of the world's nations—126 out of 183—were either democracies or were moving in that direction.

Russett identifies two sets of forces that nudge political forms in the direction of democracy. First, there are those that tend to undermine authoritarian governments. Because autocracies are not in touch with their people, they make mistakes that give rise to their own demise. An example here would be a declaration of war. Compared with democratic leaders, autocratic leaders find it relatively easy to gear up for war: they can keep their plans and preparations secret, and they can proceed without popular support. Yet each time they do this there is a chance that their government will fall as a result of military defeat or by revolution on the part of the people, who, after all, are the ones bearing the brunt of the war. The more times autocratic leaders go to war, the more likely they will put themselves out of business. Although there is a rallying effect when war is declared, war tends to decrease a leader's time in office, especially if the war is costly or is lost.[39]

Then there is the upside of democracies. This political form solves the fundamental problem of social organization at any level: providing enough regulation to coordinate different participants while leaving them enough latitude to satisfy their needs. Democracy offers a balance between the protective and facilitating benefits of government on the one hand and freedom for individuals and subgroups to pursue their own interests on the other.

The interesting possibility here is that democracy may involve a set of functional principles that will work for other intelligent species in other places and at other times. If this is so, then perhaps they will be democratic too. Francis Fukuyama believes that the flaws inherent in monarchies, fascism, and other rivals to democracy will result in their demise, leaving liberal democracy as the sole remaining governmental form.[40] This, in turn, will result in the end of ideological struggles, a major cause of wars. The ascendance of liberal democracies among so many peoples suggests that the principles of liberty and equality on which they are based are not accidents or the results of ethnocentric prejudice but reflections of the fundamental viability of the democratic political form.

Of course, there is always the chance that an extraterrestrial society will be governed by a dictator, benevolent or otherwise. Although Genghis Khan has disappeared from the scene, dictators such as Mussolini, Hitler, and Stalin were of recent advent. The awesome power and terror of their rule, their oppression of their followers and neighbors, and their nearness to us in time may color our thinking about extraterrestrial political systems. But the reality is that despite their mark on our recent past, each of these major 20th-century dictators ultimately failed and typically did so in less than the average span of one human life. Hitler's Thousand-Year Reich survived just over a dozen years. Mussolini's reign was only a little longer. Just a few short years after Stalin's "stroke," Khrushchev undertook a program of reform, and less than 75 years after the Russian revolution the Soviet Union dissolved. Other recent dictators such as Idi Amin, Castro, Peron, and Pinochet have controlled relatively small societies and have not been held in high regard by their larger, more powerful neighbors. The reality is that as history has moved forward, dictators are finding it difficult to maintain tight control over a society and many of them have already been deposed by irate citizens, offended neighbors, or both. Furthermore, most 20th century dictators have come to power in hard economic times, something that becomes less likely when the society has a large resource base and advanced technology.

Information Export and SETI

Under the SETI-contact scenario, it is our information processing at the societal (or supranational) level that will allow us to detect them. As noted in chapter 2, one possibility is that we will be attracted by a deliberate beacon; the other is that we will overhear internal communications. The former signal is likely to be statistically rare, powerful, and take the potential audience's limitations into account. The latter is likely to be relatively weak, statistically frequent, and geared for internal consumption. Deliberate signals could create very different impressions from internal communications, as we can glean from comparing local television fare with that broadcast for consumption by foreigners.

SETI is based on the assumption that extraterrestrials will export information by means of a radio beacon and that, if we are lucky, the beam will carry encoded information we can decipher. But why should an advanced alien civilization choose to share its experiences, its values, its technology? Many nations on Earth have powerful broadcasting stations to export elements of their culture and build support among foreigners. Perhaps such efforts to propagate belief systems and to win over other people represent reproduction at the societal level. Those who listen to alien broadcasts and adopt alien ideologies and technologies would not reproduce the entire alien civilization, but they would help perpetuate some of that culture's elements. Indeed, one might argue that the long-term mixing of cultural elements, like the mixing of genes, could lead to the natural selection and evolution of societies.

David Blair argues that because of the infrastructure and energy costs associated with beaming messages many light-years in many directions, extraterrestrial astronomers are likely to look for signs of life before activating a beacon.[41] This monitoring could be done optically—so, for example, their astronomers could have detected certain signs of our civilization as it emerged along the Nile—but the sensing device would have to be so large (measured in kilometers) that it lacked practicality. A much more feasible approach is to look for radio leakage. The required equipment is feasible enough, but extraterrestrials won't be able to monitor all of the skies all of the time. What they may do is look in our direction and, in the absence of any radio emissions, wait another 200 years or so to give our technology time to develop. Consequently, we should not expect to hear from them until they have detected our radio presence. If they are able to detect powerful radar signals sent in 1945, we would receive a response from a civilization 100 light-years away in the year 2145, providing that they happened to be listening in our direction when our very first radio emissions arrived at their planet. By the same line of reasoning it would not be until the year 151945 that we would hear a response from a galaxy 75,000 light-years away.

CONCLUSION

As we consider the nature of extraterrestrial life we must extend our view beyond individual organisms and consider their societies and cultures. Societies, like organisms, have observable properties, structures, and processes. The same general strategies that help educate our guesses about extraterrestrial life-forms should help us make reasonable guesses about extraterrestrial societies. We look for general principles that apply across different species, people, and historical periods as we also consider the requirements imposed by the microwave search. When we do this, we conclude that the first extraterrestrial society we discover is likely to be old, large, wealthy, and wise rather than young, small, poor, and inexperienced. Extrapolation from our experiences on Earth suggests that, despite their attraction for some science fiction writers, monarchies and dictatorships are less likely than more democratic forms of government.

How many societies last long enough to acquire immense resources, limitless information-processing power, peace, and prosperity? The only evidence we have bearing on this question is that we ourselves are still around. Some observers believe that our worries about our society's durability are greatly exaggerated. John Mueller argues that since the end of World War II, the nuclear superpowers—the United States and the Soviet Union—never came close to a nuclear confrontation. During the Cuban missile crisis, the absolute zenith of cold-war tensions, both Kennedy and Khrushchev were ready to back down rather than use nuclear weapons. Indirect confrontations between the superpowers, such as in Korea and Vietnam, were limited both in scope (objectives) and in methods (weaponry).[42]

Adrian Berry argues that many of the other threats that worry us are harmless or at least will not lead to our demise. For example, according to Berry's analysis, population growth has not occurred gradually, but in large increments when we have been able to support it. Thus, abrupt population growth occurred with the Industrial Revolution and westward migration, and the next spurt is likely to occur when we move into space. In other words, population growth may be more a consequence of our movement into space than a cause of it. As we recite the familiar litany of woes that beset our society, we may be surrendering to some form of national hypochondriasis.[43]

Given the presumed level of extraterrestrial technology, most of the reasons advanced for fearing aliens (the occupation of our planet, slavery, the plundering of Earth's resources) are implausible. It is difficult to believe that a society sufficiently advanced to support interstellar travel (a requirement for occupying our planet) would be unable to satisfy its basic resource requirements. The same Drake equation (see chapter 1) that suggests a million civilizations in one galaxy suggests many times that number of planets bearing bountiful resources but not intelligent life itself. These resources would be free for the taking, and their exploitation would not require conflict with indigenous populations such as ours.

SEVEN

SUPRANATIONAL SYSTEMS

The Vedic worldview of 8th- to 12th-century India envisioned a universal government in charge of a hierarchy of planets encompassing 400,000 humanlike races and 8 million other life-forms. The topmost authority was Brahma, who lived in a planetary system called Brahmloka. Beneath this were the systems of Tapoloka, Janaloka, and Maharloka, which were inhabited by sages, and lower yet (but still above Earth) was the realm of Svargaloka, where Devas prevailed. The humanoid inhabitants of these systems were hierarchically organized and engaged in politics and warfare. Sometimes their activities affected life on Earth. "One striking feature of Vedic accounts," notes Richard L. Thompson, "is that different races are often described as living and working together cooperatively, even though they differ greatly in customs and appearances ..."[1]

The metaphysics of early Indian religions differs in many crucial ways from that of today's science, and certainly references to "planetary systems" should be taken with a particularly large grain of salt. However, the idea of many species getting along with one another should strike a resonant note, as we have already encountered the notion that some extraterrestrial societies have already communicated with one another and have formed an association, or "Galactic Club." Depending on such factors as distances (interplanetary or interstellar) and propulsion systems, the Galactic Club could consist of communications networks or even support commerce analogous to that found among countries on different continents of Earth.

SUPRANATIONAL AND OTHER INTERSTATE SYSTEMS

Groups of societies constitute Living System Theory's highest system level. Even as we might expect to find a wide range of societies scattered throughout the galaxy, we might expect to find varied international (or interstate) arrangements. For our purposes, it is useful to distinguish among (1) isolates and independents, (2) alliances and communities, (3) empires, and (4) transnational or supranational systems. The distinction rests on the locus of the "deciders," that is, the governing or decision-making bodies.

Isolates and independents would not be true members of the Galactic Club. An *isolate* is a society that has not entered into a continuing dialogue with other civilizations. One possible reason is that it lacks the technological sophistication to receive messages across interstellar distances. Another is that it has preliminary evidence that other societies exist, but for various reasons (such as a fear of the unknown) has chosen not to pursue the matter. Or perhaps the isolate has been detected by other societies, but because it is perceived as too immature, unstable, or bellicose, they do not yet welcome it. Whether due to ignorance, choice, or rejection, Earth and other isolates play little if any role in interstellar affairs. Isolates' decisions are made internally, separate from the decisions made by societies elsewhere in the galaxy.

An *independent* society is both aware of and known to others in the universe and may even maintain open communication with them. However, it retains full autonomy and discretion. Its awareness of other societies may influence some of its decisions, but its choices are autonomous and self-oriented.

An *alliance, coalition,* or *community* exists when two or more societies coordinate their efforts for mutual gain. In general parlance, alliances and coalitions are for defense purposes and communities are for economic purposes, but there are a number of functional areas where different societies might coordinate with one another. The important consideration is that each society retains its own government and makes its own decisions, even though these decisions may be constrained or shaped in part by the choices of its partners.

An *empire* is built around a principal, mother, or "home" society that controls the foreign (and, oftentimes, domestic) affairs not only of itself but also of all the other societies that make up the empire. However, it typically delegates less consequential decisions to the individual member states. The key decision-makers are located in the home society (for example, in England when the sovereign was also emperor or empress of India) and representatives (e.g., the viceroy, governor-general) hold key positions throughout the empire. Because of the disproportionate weight given to the home society's decisions, some people view empires as large societies rather than as multistate systems.

The past century or so has seen a rapid decline, perhaps the elimination, of empires, with some of the smallest empires (e.g., Portugal) among the last to disappear. One possibility is that decline represents nothing more than random fluctuation. Another possibility is that there remain *de facto* empires, even though they may be ruled by the U.S. Congress rather than an acknowledged emperor or czar. Still another possibility is that empires contain the seeds of their own destruction. One such seed is the ruinous costs of preserving and, especially, expanding an empire. This includes both the direct costs of fighting wars and the failure to spend enough on manufacturing and other activities that build a strong, durable economy. Revolutionary tendencies in those who are governed also work against the long-term survival of empires. Colonials or natives discover that their needs are unmet, strive for independence, and often succeed.

A supranational system is an interstate system of a very particular type. It consists of societies that are under the control of a higher-order echelon. This arrangement is distinct from the alliance, where each member negotiates and barters with the others but ultimately makes its own decisions, and from the empire, where one member plays the music and the others dance to the tune.

Boundaries

On Earth, a group of adjacent states may form a supranational system, and in that case its boundaries would be easy to identify. However, societies that form associations are not always contiguous: they may be separated by oceans or by interpositioned states that elect not to take part. Similarly, it is tempting to imagine a cluster of contiguous planets surrounded by military space stations intended to prevent incursions. But here, too, there could be gaps, with some interspersed nonparticipating societies and some participating outposts located far from everybody else. Thus, while the boundaries of a supranational system may very well have some geographic features, the sharper lines are likely to be political, legal, and psychological.

At the supranational level, boundaries are maintained by the same types of institutions that enforce boundaries at the societal level, except that they are under the control of the higher, supranational echelon. These include military installations and patrols, customs and immigrations services, censors, and any other subsystems that regulate the flow of matter, energy, information, and personnel into and out of the system.

Formation of Supranational Systems

On Earth, supranational systems evolved only recently, and they tend to be piecemeal and poorly integrated, exerting only spotty control over member states. Examples of supranational systems include the United Nations, the European Union, and most other international federations. The higher echelon may extend broad control over member states, or influence only a limited number of activities (for example, warfare, international trade, environmental protection, radio broadcasting). Membership may require a massive commitment of resources and a loss of autonomy, or it may involve nothing more than abiding by mutually agreed-upon rules. A galactic union controlling trade among all the societies in a galaxy would constitute a supranational system, but so would an interplanetary association of astronomers sworn to follow standardized observational and reporting procedures.

An interplanetary supranational system would originate when two technologically advanced societies that happened to be in proximity learned of each other's existence. Given the bright prospects for life in the universe, we might expect many such pairs to form over time. Third, fourth, and fifth members would be added, and then clusters would discover each other. This would result in a pattern of slow growth and aggregation, punctuated by large increments in size as various supranational systems came into contact with one another. Since stars are much closer together in the center of the galaxy, a supranational association would be likely to start there and move outward.[2]

Supranational systems are built around spheres of cooperation. These are areas in which each member society has a vested interest; in return for surrendering some discretion and accepting limits, the individual member is assured that other societies will do the same. As James Lee Ray points out, the global community is a real community "because of a set of interrelated problems, constantly threatening to reach crisis proportions, [which remind us] that we are all passengers on the spaceship *Earth*." Problems like the population explosion, starvation, hunger, and the like, Ray adds, cannot be solved on a state-by-state basis and require higher-level or coordinated action.[3]

Growth is based on attracting new members and extending into new functional areas; for example, charters that discourage the use of nuclear propulsion could be broadened to cover other forms of pollution. Over time, more and more activities come under the control of the supranational government, so that it ends up running almost everything. The potential size of such a supranational system is hinted at by the number of technologically advanced civilizations in a galaxy. If, in a galaxy the size of the Milky Way, eligible societies were on average a "mere" 30 light-years apart, it would have 300 million potential members—more societies than the United States has people.[4]

What are the functional areas that might be encompassed by a Galactic Club? Certainly, arms control and conflict containment are possibilities. Other plausible areas include the regulation of transportation and communication, two areas where societies are likely to interfere with one another in the absence of regulation. Also salient is environmental protection. Possible concerns include toxic and radioactive waste and the accumulation of space junk over millennia of spacefaring. Pertinent, too, is the depletion of rare resources such as heavy metals. Health maintenance could be another mutual concern, particularly if disease is spread by bacteria traveling on comets.[5] In a related vein, Galactic Clubs may establish mechanisms to warn member societies about potentially cataclysmic astronomical events such as approaching meteors. We cannot be sure of the exact issues, but there are many places where cooperation among extraterrestrial societies would help each member society flourish.

As George Seielstad notes, the universe rests on a succession of processes that are based on processes quicker than themselves.[6] As the largest living system, the processes of the supranational system are necessarily the slowest of all. Furthermore, if it takes light-years for a message to travel from one member civilization to another, it is unlikely that the pace of social and political change will be rapid. Supranational systems should be quite durable in that the loss of any single member society is likely to be isolated. Interstellar or intergalactic distances would protect some member civilizations from the disasters that befall others.

MATTER–ENERGY PROCESSING

At the supranational level, matter–energy subsystems involve the conservation, development, or exploitation of resources of two or more societies. Matter–energy processing at the supranational level differs from that at the societal level in terms of control (it emanates from a higher echelon) and scale (more matter and energy are involved).

Supranational systems are described by the aggregated characteristics of the constituent societies. For example, a Galactic Club may be described in terms of the volume and shape of space it occupies; the number, size, density, and configuration of its constituent societies; the aggregated resource base (matter, energy, information); and so forth. The resources, of course, are expected to be enormous, and the expectation of an immensely expanded resource base is one of the prods that move humankind into space.

Through the sharing of information, two or more societies may develop a technology or a product that is beyond the reach of either individual society. For example, a supranational system could have an edge when it comes to unlocking the secret of interstellar travel. First, the pooled scientific knowledge of highly advanced beings from a wide range of ecological niches and cultures would be brought to bear on the problem. Second, members of a Galactic Club might be highly motivated to succeed. It is one thing to build a starship for the aimless exploration of an empty neighborhood and another thing to develop a transportation system to a specific destination. In effect, members of a supranational system already know that someone is home, and that it would be worthwhile to undertake the trip!

Given the tremendous amounts of raw materials and energy available to a Galactic Club, it is tempting to focus on megaprojects: reenergizing a dying sun, disassembling an asteroid, terraforming a planet, building a worldship, or perhaps domesticating all the energy in a galaxy to produce one of Karadashev's Type III civilizations. Certainly, such projects may occur. Yet, humans are already switching away from material-intensive products and processes and toward information-based economics.

INFORMATION PROCESSING

Information processing within the supranational system parallels information processing within the society, except that it involves larger amounts, distances, and scales. Members of an interstellar community are likely to be sophisticated information processors (otherwise they could not find and communicate with one another), and a tremendous volume of information processing would be required to inform and coordinate the member states.

A crucial issue for supranational systems is the speed with which constituent societies can communicate with one another. Although the supranational system might evolve over centuries or millennia, and although organisms that live for thousands of years may be less in a rush than we are, there may nevertheless be times when the rapid exchange of information will be useful if not crucial. Consequently, we might expect that highly coherent or well-integrated societies located within a few light-years of one another will have some means, presently unavailable to us, for the rapid sharing of information.

One possibility for reducing interstellar communication time, suggested by Claudio Maccone, is to decrease the distance that the message must be sent.[7] He proposes doing this by creating wormholes that serve as shortcuts between the two communicating civilizations. An alternative would be to increase the speed of the message. Radio waves already travel at the speed of light, and even as we do not expect to be able to travel beyond the speed of light, so we do not know how to send messages faster than light.[8] Yet, according to Donald Tarter, it may be possible to take advantage of quantum connections to beat that speed. Whereas terrestrial researchers are only beginning to conduct research on trapping, controlling, and measuring quantum phenomena, instant or real-time information exchange may not be beyond the grasp of more advanced societies. Of course, if many advanced societies have moved beyond the age of radio communication, an understanding of quantum communication would be of great help in our own search for extraterrestrial messages.[9]

Information and Trade

The chief commodity in interstellar trade, suggests Bracewell, will be information,[10] and the Galactic Club has been defined as the "central nervous system of the galaxy."[11] Indeed, the expectation is that one of the primary benefits of membership would be access to valuable technical information.[12] Virtual reality could provide the opportunity to vicariously visit other member civilizations, to learn about them and to enjoy their arts, entertainment, and tourist attractions.[13]

The ability to trade information through advanced communication and information technology may make the inability to share matter and energy inconsequential, at least if each member society has sufficient raw materials within its own solar system. Specifically, if the raw materials are available locally, advanced communications systems could make it possible for each member to reproduce the others' life-forms and artifacts. Through genetic engineering, it might be possible for one member state to populate its botanical gardens or zoos with life-forms indistinguishable from those of another world; with proper instruction, local artisans, engineers, and nanotechnologists could replicate each others' historical and contemporary artifacts. To be sure, the replicas might lack the glamour of the originals, but from a functional point of view, the replicas and the originals could be indistinguishable. Societies a million years beyond our own might be able to send instructions to create a fully preprogrammed emulated member of their own species with credentials as an ambassador.

Supranational systems tend to involve many languages, with the result that translators or interpreters are in great demand. Formally or informally, one language tends to predominate in a supranational system. Thus, German was the language of science, French is the language of diplomacy, and in contemporary times English is essential for many purposes. Although part of the supranational system, societies whose premier language is not widely known are likely to be at a disadvantage. It is unlikely that a contemporary scientist who lacks English will proceed far. Although we can't guess the specifics, we might expect that one language or communications system will predominate within the interstellar tower of Babel.

Supranational Government

Philospher John Locke viewed people as peaceful and altruistic. He argued that governments are formed partly to assure peace and tranquility, but mostly to facilitate cooperative endeavors.[14] According to Clyde Wilcox, science fiction writers who follow Locke's tradition

> see aliens as friendly, intelligent beings who voluntarily have entered into a contract for galactic and even intergalactic government with benign laws and institutions. This Lockean vision portrays galactic government as useful in guiding collective action among different species and in ensuring the

> natural rights of those species. Because the aliens are basically good, however, government sanctions are rarely needed or applied. Instead, government serves to facilitate cooperation. Lockean galactic governments are usually limited in their powers and are tolerant of the divergent cultures of the various species.*

For centuries, political scientists and their precursors have debated different strategies for doing well at international affairs. One view, advanced by the "idealists," appears to have Lockean origins. Idealists contend that people are basically cooperative and that the long-term interests of each individual society or state are best served by peaceful coexistence. Governments should refrain from aggression, and they should enter into defensive pacts. The opposing view, advanced by the "realists," perpetuates a more egotistical and aggressive approach. Realists consider it in the best interests of individual societies to take those actions that will benefit themselves, independent of the consequences for everyone else. The smart strategy for the large and powerful players in the universe is to take what they can from their smaller, less powerful neighbors. Alliances and coalitions, such as exist, are based on opportunity and a temporary convergence of interests. The realist view is consistent with simplistic notions about "survival of the fittest" and our understanding of the evolutionary advantages of size and aggressiveness.

The underlying issues is, who will prosper: collaborative societies that seek collective security and cooperate in the pursuit of common goals, or egotistical states that aggressively pursue their own ends? If cooperative states that are willing to guarantee each others' peace and freedom are more likely to survive, we might hope, following contact, to be welcomed into a community of nations. If, on the other hand, self-oriented states have the highest likelihood of enduring, we will have to be very wary of our new acquaintances. Drawing on our experience thus far, we have three causes for optimism.

The ascendence of liberal democracy (described in the preceding chapter) is important because internal political forms help

*Clyde Wilcox, "Governing Galactic Civilization: Hobbes and Locke in Outer Space," *Extrapolation*, **32**(2), 112 (1991). Copyright 1991 by Kent State University Press; reprinted with permission of the publisher.

shape foreign affairs. On Earth, the same procedures that governments use to maintain internal order come to the fore when they deal with other governments. Autocracies, which use threats and strong-arm tactics to keep the populace in line, tend toward the same approach when dealing with other nations. Democracies, which prefer bargaining, negotiation, and other peaceful means for maintaining internal social order, tend to favor peaceful solutions to international disputes. This is clear in Bruce Russett's analysis of who goes to war.[15]

Russett defines war as large-scale, institutionally organized violence involving a minimum of 1,000 battle fatalities. The number of deaths—about what we would expect in a week on U.S. highways—is meant to discount accidents, unauthorized actions by local commanders, and shows of force that are not intended to escalate to a higher level.

Let us assume that each terrestrial society has the opportunity to go to war with every other terrestrial society. In the 40-year period from 1946 to 1986, there were 28,367 dyads, or pairs of societies; that is, 28,367 chances to go to war. Russett counted 41 threats of force, 120 displays of force, 521 applications of force, and 32 wars. In other words, despite impressions to the contrary, for 40 years after the end of World War II, only 32 conflicts escalated to the large-scale, institutionalized violence of war. For each dyad that went to war during this interval, 887 dyads remained at peace.

At each level of belligerence, the dyads consisting of two democracies were less likely to escalate the conflict than the dyads that included at least one autocracy. None of the dyads involving two democracies actually went to war; all 32 wars involved at least one autocratic government. Furthermore, Russett claims, when we look across a broad range of cultures going back to the ancient Greek city-states, the same general relationship prevails, if imperfectly.

This does not mean that democracies never go to war, only that they do not go to war with each other. Democracies are aware that other democracies share their preferences for peaceful, negotiated solutions to disputes, and it is to these strategies that two democracies turn. Democracies are also aware that autocracies do not share their principles, and democracies will go to war with authoritarian states.

The principle that democracies do not go to war with each other has an important implication. The greater the number of democracies on this planet, the greater the zone of world peace. By extension, the greater the number of democracies in a galaxy, the greater the zone of galactic peace. Members of the Galactic Club should do everything they can to promote the evolution of stable democracies, because by so doing they increase their zone of peace.

Further encouragement is found in John Keegan's *History of Warfare*.[16] A British military historian, Keegan has spent his life studying warfare in all cultures and in all historical epochs; he mingles with warriors, savors their creed, explores their battle sites, and observes war's effects. He ventures that warfare, as we think of it, may be drawing to an end. In the past, rational calculations often suggested that the benefits of war outweighed its costs. In our times, marked by incredibly expensive and destructive weapons systems, the situation is reversed. A strong resource base and high technology make war too expensive. Since August 1945, not one person has died in a nuclear conflict.

Today the large and powerful states have made peace their watchword. War is less a vehicle for large, rich, and technologically advanced states to resolve their differences than a vehicle, Keegan says, for the "embittered, the dispossessed, the naked of the earth, the hungry masses yearning to breathe free, express their angers, jealousies, and pent-up urge to violence."[17] The richer nations have been shielded from most of this violence, but as media coverage increases everyone's awareness, they are trying to find ways to suppress conflicts among their less affluent neighbors.

John Mueller reaches the same general conclusion as Keegan does.[18] He suggests that the past 400 years has seen steady increases in the proportion of people who believe that war is abhorrent (repulsive, immoral, uncivilized) and methodologically ineffective (expensive and futile). The American Civil War and World War I accelerated the rejection of war among many nations. Long before the outbreak of World War II Holland, Sweden, and Switzerland had renounced war. In the 1930s, attempts to appease Hitler hint that war itself was seen to be a greater enemy than Hitler, and following World War II, the political regimes that had replaced the former Axis powers were among the foremost advocates of peace. These countries traded their former status as major powers for prosperity, moral and economic independence, and respect. From 1945 on, the increased costs associated with nuclear warfare only strengthened pacifistic tendencies.

Mueller provides much food for thought for those who believe that war is inevitable. Many countries, including former enemies, have existed side by side for decades without armed conflict. Nobody expects a resumption of hostilities between the United States and Canada or among France, England, and Germany. Mueller sees strong parallels among people's attitudes toward slavery, dueling, and warfare. At one time slavery and dueling seemed natural and justifiable. Gradually, these institutions became peculiar, then repulsive, and then faded from contemporary human experience. War, like slavery and dueling, may become "subrational," that is, given no serious thought as an option. Even as we don't think of stepping into an open elevator shaft to get to a building's basement, we won't think of war to resolve an international dispute.

One of the few positive effects of war is that it unites people, if only to overthrow a common enemy. Ian Crawford argues that an active space program is a "moral equivalent" to war, in the sense that it could unite many nations in an overall effort.[19] For the individual, spacefaring provides opportunities for meaningful risk and recognition; it will also capture the imagination of millions of people who will not participate themselves. Additionally, notes Crawford, "space exploration, like astronomy before it, provides a cosmic perspective from which to view human affairs, a perspective which makes the division of our planet into a couple of hundred warring nation-states seem ridiculous, even obscene."[20] Here is yet more hope: spacefaring societies will have a grand, cosmic perspective that will make wars seem silly.

Computer simulations of interstate dynamics allow the political scientist to control variables and to look at many "what if" scenarios. Their findings are state-independent (that is, they are not contingent on the specifications and history of specific terrestrial nation-states), and their results have broad applicability. Thomas Cusack and Richard Stoll have undertaken an ambitious research program with a computer simulation called EARTH (Exploring Alternative Realpolitik Theses), which is capable of modeling interstate dynamics involving 2 to 6,400 states.[21] The studies of particular interest here involved 98 states. Cusack and Stoll varied the number of states that entered into collective-security agreements. The idealists believe such arrangements extend the life of the individual states. The realists believe that such arrangements won't work unless all societies enter into the agreement, that not all of those that enter into the agreement will abide by it, and that such arrangements could even decrease the expected life span of a society or nation-state.

The project involved 3,240 computer simulations reflecting many combinations of variables. These included the external-power orientation, the amount of power, the ability to estimate one's own power and that of other states, the internal growth rate, the cost of war, and the degree of restraint that was to be shown in the case of victory. There were four external-power orientations (foreign policies). The two orientations that stood in the sharpest contrast were primitive power seekers and collective security seekers. *Primitive power-seekers* initiate disputes if they are more powerful than their neighbor, select the weakest neighbor as the target, join offensive alliances if those are more powerful than defensive alliances, and join defensive alliances only if these are more powerful than offensive alliances. *Collective-security seekers* never initiate disputes, never join offensive alliances, but always join defensive alliances. Because history is ongoing, the computer runs involved hundreds of iterations, or "plays." That is, the ending point of one simulation would serve as the beginning point for the next. Because the results yielded 317,520 individual state histories, data analysis was based on a sample.

The simulation showed that individual states practicing collective security survive longer than states following other foreign policies. On the whole, interstate systems containing states with a collective-security orientation survived longer than those that did not. Contrary to the realists' expectations, it was not necessary for all members of the system to join the collective-security arrangement. The more states that belonged, the longer the endurance of the system as a whole. "Our results," state Cusack and Stoll, "suggest that practitioners of collective security are ecologically superior to states following other, more self-interested and clearly realist strategies."[22]

These studies do not require us to rely on world history for forecasting interstate systems in other locations. On the whole, there's hope that if paranoid, berserk, and thoroughly selfish societies last long enough to make contact with other civilizations, their foreign policies put them out of business.

BUT WHERE ARE THEY?

SETI is based on the belief that our galaxy is liberally sprinkled with civilizations older than our own. At least some of these societies should be spacefaring, and some should have undertaken space colonization long ago. Over the ensuing hundreds of thousands or perhaps millions of years, successive neighboring planets and then the solar system would become colonized. Given the copious amounts of time available to the front-runners in the race, we might expect extraterrestrial life-forms to be widely distributed throughout our galaxy.

Yet if countless civilizations are out there, why have we yet to see evidence of even one of them? If a Galactic Club exists, why are we still waiting for an invitation to join? This question, typically known as the Fermi paradox because physicist Enrico Fermi was one of the first to raise it, is far from trivial. The most obvious answer—we have not seen them because they do not exist—has given many otherwise optimistic people pause.

Perhaps the first person to write on this issue was a Russian, Konstantin Tsiolkovsky, who did so during the early decades of the 20th century. Tsiolkovsky, one of the most profound thinkers on spaceflight and a staunch believer in extraterrestrial intelligence, was troubled by the absence of evidence of extraterrestrial life.[23] One possibility that occurred to him was that they probably will visit us, but that it was not yet time. Many thousands of years passed, he noted, before the Australian aborigines and the American Indians were finally visited by Europeans. He also thought that we lacked the means to identify the extraterrestrials' presence in the universe. But the thrust of his thinking was that we are not yet ready for higher beings to contact us: "We are brothers but we kill each other, start wars, and treat animals brutally. How would we treat absolute strangers?"[24] He believed that contact could ruin us, and, because they are aware of this, older and wiser civilizations leave us alone. Another possibility that occurred to Tsiolkovsky was that "we have been set aside as a reserve of intelligence in order to allow our species to evolve to perfection and thereby bring something unique to the cosmic community."[25]

Over the years, many people have sought to reconcile the belief that extraterrestrial life exists with the fact that we have yet to encounter them. Robert Freitas suggests that the issue can be expressed as a syllogism: (1) If extraterrestrial civilizations exist, then they have probably spread out across the universe by now. (2) If they have populated the universe, then we should have seen some evidence of them by now. (3) Since we have not seen them, they probably do not exist. He then attacks this reasoning on logical and evidentiary grounds. The logical problems reside in the use of probability statements as the logical operators (Freitas points out that the argument is framed in terms of "probably," "ought to be," "should be," "might be," "hoped to be," "likely," and "reasonably," rather than in terms of "is"). The evidentiary problems have to do with the assertion that "we have not yet seen them." Perhaps we have misinterpreted evidence or perhaps we have not yet taken a close enough look.[26]

Would They Expand throughout the Galaxy?

Many researchers, operating with varying assumptions, have concluded that ET has had plenty of time to hop from star to star and should be in our solar system by now. Frank Tipler, for one, has estimated the speed with which civilizations might expand throughout the Milky Way galaxy,[27] drawing on John von Neumann's concept of the self-replicating automaton. The basic idea is that machines would be constructed to explore nearby solar systems. These machines, requiring a substantial initial investment, would be able to duplicate themselves and hence multiply as they proceeded with their scouting missions. Beyond the initial fleet, there would be no cost because the automatons would use local resources to build clones of themselves. So, for example, an automaton might proceed 4 light-years to the nearest star, pause for a few centuries, build two more exactly like itself, which would then go to the next two stars. The two automatons would become four; the four, eight; the eight, 16; and so forth in geometric progression. Martyn Fogg calculates that not only has there been enough time for intelligent life-forms to spread throughout our galaxy, but also sufficient time for it to spread throughout the entire universe.[28]

Perhaps resource limitations inhibit expansion. Perhaps even given a giant resource base, interstellar expansion isn't worth a society's investment. On the other hand, there may be ways to keep the cost down. For example, a society could settle for slow ships, rather than fast ships (see chapter 2), and the galaxy could be populated in a succession of waves rather than overnight. Furthermore, as previously noted, probes sent forth to colonize the galaxy could be very small, perhaps even the size of a fist or virus. All that is required is an information code and an ability to synthesize local materials. The energy costs could be contained by using a small probe proceeding at slow interstellar speeds.

Given a suitable industrial and technological base, would there be motivation to undertake galactic colonization? M. D. Nussinov and V. I. Maron are not convinced that such motivation exists. Advanced civilizations probably have abundant resources in their immediate neighborhood. They are also likely to have immense information-processing systems that provide them with all the knowledge they need to prosper. "So," these authors conclude, "there would be no impelling reasons for search and contacts with other civilizations, because such attempts would require a lot of energy and time as well as serious efforts."[29] On Earth, not every culture is interested in exploration. The Tibetans, for example, were amazed by the Europeans' drive to conquer Mount Everest. British explorers were seen as rich lunatics who were locked into a foolish quest. They were viewed, states Neil Steinberg, "somewhat the way New Yorkers would view a Japanese billionaire who arrived in the City and announced it was his mission to lick the surface of the Chrysler building."[30]

Even if an extraterrestrial society had both the means and the will for colonizing our galaxy, there are several ways our solar system might have somehow been missed. Perhaps there was a break in the geometric expansion (the probe designated for our solar system may have run into problems along the way). Maybe they will arrive as I finish writing this sentence. The Milky Way galaxy could be in large part colonized, but there could still be pockets of isolates such as ourselves.

Although Tsiolkovsky's work was not well known, several hypotheses, developed independently in more recent times, share his theme that intelligent extraterrestrial species may be nearby but for various reasons prefer not to make themselves known. One variation contends that Earth is being monitored for its levels of technological and social development, and that we have not yet proved ourselves worthy of recognition by the mature and responsible societies of the universe. Another variation is the "embargo hypothesis," which states that it is unethical to interfere in the workings of a developing society.[31] Related, also, is the possibility that we are under scientific study, and that even as anthropologists would strive not to influence the cultures they study, aliens would prefer to keep us in our natural state. Because they are remaining unobstrusive, they are unseen. Quite reminiscent of Tsiolkovsky's ideas is John A. Ball's Zoo hypothesis, which suggest that we are being maintained as a natural preserve, much as a zookeeper might try to maintain a natural preserve for animals.[32] Within our preserve, we are valuable for amusement purposes, if not for serious study. (One might wonder how the aliens would maintain cover. If I were conducting such a study "in person," I would arrive on Earth disguised as an oldster or as a child. Compared with adolescents and adults, old people and children are allowed to come and go and, in the absence of a fuss, tend to be ignored!)

Michael Hart proposes that no sociological explanations are adequate because any such explanation must account for all civilizations at all times.[33] Given the large number of civilizations expected by many SETI enthusiasts, we might expect exceptions to any sociological principle. For example, although there might be an almost overwhelming tendency for maturing civilizations to shift from outward expansion to inner contemplation, there must be at least one interested in colonizing the galaxy. Or even if 99 percent of the civilizations that arise find it too expensive to migrate in our direction, at least one would have the resources to do so. Or, whereas ethical principles may keep virtually all societies from impinging on one another's territories, shouldn't at least one unscrupulous society be eager to do so?

The problem with this seemingly self-evident analysis is its presupposition that individual societies are free agents that act independently, without any constraints or obligations imposed by other societies or a supranational system. Interstate dynamics limit what any individual society may choose to do. Societies run into opposition from other societies, from alliances and coalitions, and from organizations of societies, such as NATO or the UN. Despite enormous wealth and power a society might not be free to migrate to our solar system because other societies, alone or in combination, keep it out. For example, among the approximately 180 nations on Earth, at least one of them should want to take over a small, rich society like Kuwait. In fact, one nation (Iraq) did seek to do so, but in the final analysis its bid was foiled by a coalition of other states. Analogously, it is not necessary for all extraterrestrial societies to voluntarily refrain from entering our solar system. All that would be required is a policy of noninterference on the part of a supranational system (or at least a dominant coalition) that exerts control over events in our part of the galaxy.

Poul Anderson advances the theory that we could expect an alien society to undertake a limited investigation of the galaxy, but that it would have no compelling reason to explore every nook and cranny.[34] Let's assume that a very old civilization has already made contact with 50,000 other civilizations. The motivation to find yet more civilizations would be slim. Beyond a certain point, the exploring civilization would be so saturated with information that to continue would be pointless, and this saturation level could be attained prior to contact with Earth.

Anderson's argument must make great sense to any scientist who has far more data than he or she could ever hope to analyze. Scientists do not attempt to explore every instance; this is inefficient and unnecessary. Instead, we sample. If we want to determine impurities in a water supply, we analyze a few tankfuls because it is inefficient and unnecessary to analyze the entire reservoir. This is true even in the case of an exact science such as astronomy. If anyone ever actually sat down and counted the number of stars in the Milky Way, it is unlikely that the result would be rounded off to twenty-odd places. Instead, astronomers do well enough by sampling areas of the sky and then extrapolating to the sky in its entirety. The Cusack and Stoll study on the stability of interstate systems involved hundreds of thousands of state histories. The conclusions that collective-security arrangements conferred an advantage were based on a 1 percent sample. Rather than analyze the entire set of results they were better off sampling and spending the rest of their time in more fruitful activity.[35]

Can We Conclude That There Is No Evidence?

Whereas we terrestrial observers may have no evidence of ET's presence, this lack may reflect nothing more than shortcomings in our observational powers. There are three themes here: we have the evidence right under our noses but do not recognize it; the evidence exists but we haven't looked hard enough to find it; the evidence has been found but the government is hiding it from the public.

The theme that the evidence is before us but that we are ignoring or misinterpreting it returns us to the idea of UFOs as visitors from outer space, and to "ancient astronaut" theories that extraterrestrials visited Earth, maybe bringing life here millions of years ago (in which case we ourselves could be part of the evidence), maybe visiting us between 3,000 and 5,000 years ago, maybe more recently. In support of this, it is argued that Earth's own ancient civilizations made rapid progress in certain areas and were able to develop certain artifacts such as pyramids only through the intercession of advanced outsiders. Also relevant are stories, including detailed physical descriptions and even drawings, of unusual beings who came to provide enlightenment in food production, government, and culture as well as engineering technology.[36] Such "evidence" tends to dissipate on close examination; otherwise, the question Where are they? would be, at best, a historical footnote in the (resolved) debate about intelligent life on other worlds.

We should attend, however, to the view that we haven't looked hard enough to conclude that there is no evidence of their existence. Only recently have we entered the radio era and our microwave searches are in their infancy. In addition, probes and other artifacts may be too small to be noticed accidentally by our optical telescopes and radar sets.[37] Dedicated optical and other searches might identify interesting artifacts, but at the required level of resolution the skies are 99.99 percent unexplored. Freitas proposes using existing or foreseeable technology to continue a search for artifacts in the 1- to 10-meter range in selected orbits. He adds that there are no objects intended to attract our attention (otherwise we would have found them) and that we will not be able to detect those objects intended to elude us. Our best chances are to find objects for which detection by us is, to them, unimportant.

Last (and probably least) is the argument that clear evidence of their existence has already been found, but that this evidence is known only by people with the highest security clearances and deliberately kept secret from everybody else.[38] This possibility, like the possibility that flying saucers are extraterrestrial spaceships, captures the imagination but is given more attention by UFO buffs than by SETI scientists. So, we end up with one contingent obsessing over UFOs, abductions, the great pyramids, Stonehenge, and the face on Mars as evidence of beings in outer space, while other contingents obsess over the absence of evidence that ET exists.

CONCLUSION

Societies are not independent agents that are always free to pursue idiosyncratic, egocentric ends. Rather, they must operate within contexts set by other societies. To thrive within an interstate system, they form alliances, coalitions, communities, blocs, and supranational systems. Living Systems Theory contends that larger and larger biosocial systems have evolved over the aeons and that supranational systems are the ultimate result. These clusters or groups of societies are based on cooperation and its members have subordinated themselves to higher-level decision-makers or governments. As the largest and most complex living system, the supranational system is, in effect, the ultimate in multicellularity.

What types of relationships might we expect among extraterrestrial societies that have already encountered one another? SETI proponents suggest that extraterrestrial societies will be peaceful and cooperative; ugly societies will bring about their own downfall and rapidly disappear. This wish is very understandable—after all, who would want to invest in a search that could lead to Earth's downfall?—but terrestrial experience suggests it has a factual foundation. First, there has been a pronounced drift over the years away from monarchies and autocracies and toward democratic forms of government. This is important because internal politics shape foreign affairs. Although they go to war with autocratic societies, democracies do not go to war with each other. If, as suggested here, democracies rest on organizational principles that are of universal applicability, then we would expect many democracies and a large zone of peace.

Second, despite the enormous bloodshed of two world wars and a half century or so of living under a nuclear threat, more and more people view warfare as repugnant and ineffective, the large-scale institutionalized violence of war is on the decline, and it is possible to glimpse a world of peace. Finally, as computer models of broad applicability show, societies that refrain from exploiting one another but that rush to one another's defense are likely to outlast others. All of this suggests that despite our intuitive notions about size, aggressiveness, and the survival of the fittest, berserk or belligerent societies are likely to collapse; it is peaceful societies that we are likely to meet.

On a very practical level, interstellar distances should work against authoritarian interstate systems. Such governments require close monitoring (name an authoritarian, coercive form of government that did not require a network of police agents and informants). It would be difficult to monitor a society light-years away, and, unless the electronic detection systems were very sophisticated, monitored societies would have opportunity to falsify the data. Moreover, because of the immense distances involved, it would be very difficult and expensive to apply military force.

The problems of surveillance and enforcement, and those of simply managing a variety of species on a diversity of worlds, hint that the Galactic Club will be organized along a few broad principles. Perhaps numbering no more than a dozen, these principles would establish only general guidelines. The government of a member society would not be a puppet but, rather, gifted at such tasks as extrapolation and interpolation and applying these basic guidelines to local conditions. The true genius of leadership would consist of implementing the same fundamental principles in locations that might differ substantially from those in which they were formulated. The members of the Galactic Club will not all be alike; on the surface, especially, there will be considerable variation among the constituent nation-states.

We might expect members of a Galactic Club to be used to a level of heterogeneity and diversity far beyond our ken, and to have evolved various protocols for dealing with newcomers. One guess is that their culture will be cosmopolitan, or perhaps we should say "cosmos-politan." That is, as a Galactic Club grows, members become adroit at communicating with off-planet species. Over time there might evolve a whole set of standardized procedures for learning about and enrolling civilizations such as ours.

Several analyses suggest that there has been plenty of time for societies to spread throughout the universe. Since we have not yet discovered evidence of them, they must not exist. Therefore, SETI is a waste of time. Yet, there are too many ways to reconcile the existence of many extraterrestrial societies with the absence of evidence of even one such society. Perhaps we are overlooking the evidence; perhaps we haven't looked hard enough. Perhaps the immense distances involved and the impracticality of faster-than-light travel keep extraterrestrials at home. Despite claims to the contrary, sociological explanations are no worse than any other, particularly when we raise our sights beyond the individual society and take interstate dynamics into account. All it takes is one society intent on exploration to initiate the process of populating the galaxy. All it takes are two societies to bring this expansion to a halt. It is not necessary for all societies to be alike for a sociological explanation to hold true.

EIGHT

FIRST IMPRESSIONS

Before the American Indians and Europeans actually met, notes James Axtell, each had preconceptions as to what the "other" would be like.[1] The Indians, who were relatively isolated and whose experience was limited to other Indian peoples, were prepared to see others either as equals and pretty much like themselves, or as superiors that resembled their anthropomorphic deities. Europeans, on the other hand, could draw on centuries of experience on three continents. They were aware of people of differing sizes and shapes with black, brown, and yellow skins; political systems including anarchies, democracies, and monarchies; religious beliefs including atheism, Buddhism, Christianity, and pantheism; and all sorts of customs, habits, and languages. Moreover, for centuries Europeans had been exposed to descriptions of "strange people who ate human flesh, peered at the world through one large eye in the middle of their chest, and barked rather than spoke from canine snouts."[2] (These and many other legendary beings were "confirmed" by explorers, and 4 of the 12 best-sellers of the 14th and 15th centuries dealt with such marvels.)[3]

Although, compared with the American Indians, Europeans

were more aware of diversity in the world, people who were different from them were considered inferior. Despite a few works that heralded the wonders of the new world and waxed poetic on the nobility of the American savages, by 1530 a "negative, demonic view of Indian cultures had triumphed and its influence was seen to descend like a thick fog upon every statement officially and unofficially made on the subject."[4]

Given our tendencies to be naive, suspicious, or unreceptive to other human beings who are not "one of us," how will we view truly different intelligent life-forms? Of course, most of us don't really expect ETs to visit us in person. But a confirmed radio signal or other strong evidence of extraterrestrial intelligence will trigger a flood of questions: Who are they? Where do they come from? What are their intentions?"

Despite scant or ambiguous information and dubious facts, the human mind is capable of developing comprehensive and detailed pictures. We form complex impressions on the basis of minimal information because hard-wiring, psychological programming, and cultural conditioning help us find, organize, and interpret "the facts." Our perceptions represent a mix of two types of information: that arriving via the senses, and that originating within the self. To the extent that hard data are lacking or ambiguous, we tend either to look to other people or to look inward to find order and meaning. Since there may be little hard information at the moment of contact, our motives, needs, fantasies, expectations, and prejudices will heavily influence our perceptions of what "they" are like.

In all likelihood most of us will think about extraterrestrials in the same way that we think about other humans. As noted in chapter 1, the process of ascribing human abilities, traits, and motives to nonhuman species is called anthropomorphism. This tendency, notes British ethologist John S. Kennedy, is natural and pervasive and dinned into us from early childhood.[5] We could not abandon this tendency if we wanted to; besides, we do not want to. Most likely they will think about us in terms that are available to them, and our impressions of each other may or may not correspond to the impressions that each of us has of ourselves.

Whatever their origins, our first impressions will be critical. For many people first impressions will provide a basis for sweeping generalizations regarding their capabilities, intentions, and (from our perspective) overall worth. Accurate or inaccurate, first impressions are difficult to change, in part because we tend to look for confirming evidence and neglect information suggesting that our views are wrong.

Beyond their generality, persistence, and powerful role in determining initial reactions, first impressions initiate cycles of events that will affect the relationship between the two societies. Specifically, our expectations about how others will act tend to bring the expected performance about. This is known as a self-fulfilling prophecy. The impression that we have contacted a benevolent culture is likely to elicit an open and friendly response on our part, which is an invitation to them to respond in kind. On the other hand, an unfavorable first impression, which gives rise to fear or suspicion and causes us to withdraw or respond defensively, could prompt defensiveness on their part. Their reciprocal defensive reactions would confirm our original suspicions, thereby elevating our fears. The result is a self-amplifying or positive-feedback loop, characterized by an escalation of tension and hostility.

In this chapter we consider three factors likely to mold our initial impressions of whatever beings we discover "out there." First, there are input or message factors, that is, variables that shape the accuracy or tone of what we receive. These variables are crucial because they establish the "facts" on which our impressions will be based. Second, all but the tiniest fraction of people on Earth will not hear *from* the ETs but *about* them. Thus, most people's impressions will be based less on the evidence than on journalists' and newscasters' interpretations of the evidence, interpretations that may not square with the basic facts. Consequently, we must consider news dissemination, that is, how the release and handling of the news will affect world opinion. Finally, we will consider some of the psychological and social processes likely to affect our initial impressions.

MESSAGE FACTORS

Although we could encounter a live ET or stumble across an alien probe or artifact, the most likely scenario is that first contact will consist of our detecting extraterrestrial radio activity. As Donald Tarter points out, most discussions seem to assume that the realization that we are not alone in the universe will take the form of a sudden insight. The first signal we intercept, however, may be of ambiguous origin. Only after a long, slow process of assembling bits and pieces of information will we be able to conclude that a transmission is of intelligent extraterrestrial origin. Thus, recognition of ET's existence may take the form of a dawning realization with gradually increasing numbers of people accepting the discovery over time.[6]

If the evidence is flimsy or ambiguous, people will differ in terms of their willingness to accept claims of contact. Years ago, Mary Connors suggested that the single most important factor determining the degree of acceptance of an announcement regarding the discovery of an alien culture would be people's preexisting beliefs concerning the existence of extraterrestrial intelligence.[7]

The initial intercept is likely to be a carrier wave that by itself contains minimal information but that will, in this case, have certain formal properties suggestive of intelligent extraterrestrial origin. A second look, with different equipment, may reveal that it is indeed an uninformative transmission (a homing beacon, perhaps) or it may reveal information that is superimposed on the carrier. Under minimally informative conditions—that is, our knowing nothing about the possible content of the message—we could still make a rough determination of the level of the technology, tell whether there is information superimposed on the carrier, ascertain the direction from which it came, and estimate the distance to the point of the transmission's origin.[8] We may be able, moreover, to tell whether the point of origin is approaching us or moving away; if the extraterrestrials are broadcasting from a planet, that planet's speed of rotation.

If we discover information superimposed on the carrier wave, then we begin the process of trying to extract it. As discussed in chapter 2, profound biological and cultural differences between extraterrestrials and humans are likely to make this task daunting and frustrating. Because we expect few helpful reference points, and because extraterrestrials and humans may think in fundamentally different ways, we must be prepared for a long, tough haul that may get us nowhere, may result in our deciphering only parts of the message, or may lead to our understanding the message in its entirety. As Tarter points out, one of the most difficult scenarios, psychologically, would be if we knew for sure that the transmission was of intelligent extraterrestrial origin but had little or no idea as to what it meant. Under these conditions we will see a thriving interpretation industry.[9]

The message, of course, could be anything. As noted in chapter 2, scientists argue that the physical world offers one of the few reference points that both humans and aliens have in common, so the message itself could focus on matters that interest scientists and engineers. If this is correct, the transmission will reveal nothing about their biology, psychology, sociology, culture or politics, but will provide food for thought for physicists, mathematicians, cryptanalysts, logicians, and people who do the kind of magazine puzzles that are so arcane that most subscribers don't even understand the goal.

Even if the transmission is unambiguous and touches on many different topics, it is unlikely to give us a complete picture of the extraterrestrials, their intentions, and their posture toward us. A single message will not capture the essence of a totally foreign and ancient culture, the richness of its history, and its strengths and weaknesses, or, as a friend once commented, help us comprehend their equivalents of a Mickey Mouse birthday parade or tradings in pork-belly futures. There are many reasons why an interstellar message might be incomplete, ambiguous, or misleading—in other words, why it might be exactly the kind of message that will put a great burden on our imagination and interpretive skills.

How accurate and complete are the messages that we ourselves have sent forth via space probes? The *Pioneer 10* (and *Pioneer 11*) plaque depicts a nude human couple, the man's right arm upraised in a wave, a gesture that we hope (but do not know) is a universal symbol of peace. The nudity makes sense because it frees us from culture and historical epoch and acquaints the beholder with our biological organism. At the same time, it is only under rare conditions that we encounter one another nude; more generally we appear in public wearing clothing. A small matter, perhaps, but the expectations generated by the plaque might not be fully confirmed by firsthand observations on Earth.

Then there is the plaque on the moon stating, "We came in peace for all mankind." This one-liner is at best a half-truth. This message does not reflect East–West tensions or the war that raged in Vietnam at the time it was prepared. It is true that our exploration of space is a peaceful venture and so, perhaps, is the message's hint of an openness to any species that might find it. But it presents a selective picture, one intended to serve our purposes if it is someday discovered.

Control of the Transmission

We hope that we will hear from a society whose members speak with a unified voice, at least on critical matters, so that the message reflects their core values and the prevailing social order. Because the message is all that we will have, there will be a strong temptation to treat its contents as indicative of their culture as a whole. Unfortunately, there are alternatives to this simple state of affairs.

One possibility is that the broadcast was not under the control of one unitary government but instead one of a number of independent nation-states. As on Earth, the fact that these nation-states may coexist does not mean that they are in accord. They may have special interests, and these could shape their transmissions. Our ignorance of the variability among their governmental forms will make it difficult to interpret their messages, even as one of their observers might have trouble recognizing the slants of messages emanating from Beijing, Moscow, Washington, and Tel Aviv. Another possibility is a renegade transmission, that is, one under the control of an entity operating outside the mantle of government. Renegade broadcasters could be anything from a group of rebels intent on overthrowing the local planetary government to pirates to drunken or crazy interstellar radio hams. Yet because we are more likely to receive a strong and prolonged signal than a weak or brief one, it seems unlikely that we will hear first from an unstable group.

As noted in chapter 6, societies develop specialized subsystems that serve boundary-spanning functions; that is, the subsystems regulate the matter, energy, and information that flow in and out of the society. Such specialized subsystems, perhaps analogous to our Department of State or military, could be in charge of their transmissions. Diplomatic control of a transmission is likely to paint a very particular picture, one intended to convey a positive image and at the same time, subtly or not so subtly, create reactions that are of benefit to the broadcaster. We might expect, in effect, an enhanced Radio Moscow or Voice of America transmitting interesting information, but information that has an unfathomable but crucial spin. Military transmissions, on the other hand, may convey a state of readiness and a determination to prevail in any confrontation. Neither the diplomatic nor military versions of their world are likely to be entirely accurate, but we would expect the former to create more favorable first impressions than the latter.

Still another possibility is that we will intercept a message that is not intended to be taken seriously, the equivalent of a commercial radio or television broadcast. Imagine the level of consternation of beings on another planet that eavesdropped on the famous *War of the Worlds* broadcast, which created the impression that Earth was being invaded by Martians in 1938! Our television signals tend to be fairly weak by interstellar standards, but let's suppose that by some means our television shows are somehow detected and unscrambled. What would they say about our culture? Social psychologist Donelson Forsythe surmises that "after watching our situation comedies, police shows, detective thrillers, hospital dramas, soap operas, and commercials, they would conclude that the average human being is witty, violent, deceitful, ill, selfish, sexually driven, and vain."[10]

Elapsed Time

Due to the vast distances traveled, the message we receive could be hopelessly out of date. A society might disappear in a nuclear holocaust just years after reporting stable conditions. Consider how a contemporary description of world politics would differ from that of 50, 25, or even 10 years ago. Extraterrestrials who drew a picture of intense competition or rivalry between our Eastern and Western hemispheres might expect this state to continue after determining that one of our messages was only 10 (Earth) years old. Yet the situation has changed dramatically during that period of time: we have seen the unraveling of the Soviet Union, the unification of Germany, and so forth. Many of these changes represent a de-escalation of world tensions, but the opposite could have occurred. Whereas we recognize that major changes can occur over long periods of time, we may be less sensitive to the possibility that momentous changes may occur in very short order. Our own history suggests that change is becoming more profound and rapid.

Fortunately, there are some ways to estimate the age of an incoming message. In effect, the older the message, the less confident we can be about the current accuracy of its content. And, if we choose to respond, the further away they are, the less confident we can be that the message they receive will provide a currently accurate statement about life on Earth.

Inhibition and Guardedness

Unless they are inveterate warriors or proselytizers, they may develop a wait-and-see attitude. It makes sense not to commit oneself fully before learning about a newfound acquaintance's capabilities and intentions. (It is in part for this reason that we ourselves prefer a passive search.) If one wishes to make friends, it is useful to develop sufficient knowledge of the others' culture so as not to commit a *faux pas* or otherwise create a bad impression. Furthermore, those who gain experience in different cultures understand that the wise strategy is to observe the others' behavior and then follow suit. If the extraterrestrials are quite guarded in their own behavior or attempt to blend in with our ways, we will have relatively little basis to learn what they are like. Alternatively, we may through the process of projection conclude that they are similar to us in ways that they are not.

Deceit

Many of us expect that a highly advanced society would live in secure, comfortable surroundings and have no need to cheat, steal, or lie to survive, but this is far from guaranteed. Perhaps extraterrestrials have achieved their superiority through fraud and misrepresentation (although I have doubts that this would be successful in the long run). Perhaps their survival is based on an ability to deter interstellar interlopers by giving the impression of overwhelming military superiority or by presenting themselves as repugnant.

Deceptive behavior is widespread among Earth species, including humans. Whatever our moral judgment, deception often works to the deceiver's advantage. Predators trick their prey by appearing innocuous or by using stealth rather than a frontal assault. Prey have ways to escape their predators, such as by melting into the scenery or appearing much more ferocious than they really are. Bluff and deception have also played major roles in diplomacy and warfare on Earth, and many science fiction stories have capitalized on the role of deception in luring humans into traps.

Electronic communications are notoriously easy to fake. The supranational system or the society that controls the transmitter will have complete control over the message and can choose what it reveals. Already humans are able to construct bogus but compelling messages. This is routinely done by news establishments that can find a startling piece of information in the most mundane of situations and can control the public's response by means of careful editing. As noted in Ian I. Mitroff and Warren Bennis's *The Unreality Industry*, much of what passes as news is a blend of fact and fiction.[11] With computerized graphics it is even possible to create a picture pixel by pixel if so desired.

Apart from a few clues provided by the technical properties of the transmission itself, there are few avenues for verifying the information received over interstellar distances. Normally, when one nation receives a message from another, it has ways to determine the truth, perhaps by examining the message in light of past experiences with the communicator, perhaps by relying on satellite surveillance or reports of operatives and spies, and so forth. Under most scenarios of alien contact a radio message will be all that we (or they) have.

Thus, to some extent our first impressions will be based on "facts." But due to the nature of their transmission or our limited ability to understand it, these facts may or may not tell us almost nothing about their life-forms, personalities, politics, and other topics likely to gain the attention of the public. Furthermore, even if we rapidly uncover tantalizing detail, we must be open to the possibility that the picture that emerges does not accurately represent their society at the time we hear about it.

DISSEMINATION OF THE NEWS

Actual "contact" will involve very few people, at least at first. The first extraterrestrial transmission we intercept is likely to be identified by a small research team working at a radio astronomy facility. Even the most dramatic contact scenarios, such as an alien spacecraft hovering over Los Angeles or landing on the east lawn of the White House, will involve a small number of eyewitnesses.

Our impressions and reactions will thus depend, in large part, on how word of the contact is distributed to the world at large. Of particular interest are the kinds of deliberate or inadvertent omissions, fabrications, and distortions that could shape public reaction. At least in the short term, the actual qualities of the aliens may be less important than the stories that are circulated.[12] Error-free communication over interstellar distances is only one of our wishes; another is the accurate and responsible dissemination of the news to minimize rumor, confusion, and disbelief.

Protocols and Leaks

SETI scientists and international organizations of astronomers profess great concern about verification of the signal and release of the news. The International SETI Protocols (which state what to do once detection has occurred) prescribe great pains to avoid false reports and to ensure that information presented to the public is highly accurate,[13] and Donald Tarter has proposed a Contact Verification and Interpretation Committee to assure the orderly dissemination of accurate news.[14] We should applaud this sensible concern about the release of the information to the public.

But how likely is it that the protocols will be followed? It will be very difficult to maintain secrecy pending an official announcement. Too many organizations and too many people will be involved for secrecy to be maintained, particularly in this era of e-mail. Quite a few people, by virtue of their position or associations, may get word that something momentous is afoot. What they learn is likely to be incomplete if not misleading.

History offers an encouraging precedent. People involved in the Manhattan Project (which developed the first atomic bombs) seemed to do a fine job of keeping their mouths shut. However, conditions were different then. There was a war on, and the project was under strict military control. There was a massive publicity campaign to discourage loose talk of any kind. Wartime censorship was taken as a necessity, not as a violation of the right to free speech. Occasionally, there was news that, if published, might hint of the development of atomic weaponry. The media, however, honored the government's requests for a complete blackout on news relating to the topic.

A less encouraging precedent than the Manhattan Project is the recent history of the politics of science. Critics argue that contemporary scientists have been corrupted by the need to obtain grant money at all costs and by the requirement of an endless parade of discoveries for professional advancement. This has led to grandstanding, bickering, and the occasional accusation of fraud. (The public altercations over who discovered the AIDS virus and possible advances in AIDS immunology illustrate this.) Perhaps we no longer have a scientific community united in zealous research, but rather various camps intent on establishing the primacy and genius of their own work. Such bickering would not serve us well in our postdetection attempts to inform and enlighten the public.

After contact but prior to the media's dissemination of the news, word is likely to spread among scientists and politicians. These are people who will have contacts with the "insiders," but who are not supposed to receive notice in advance of the official news release. The trusted recipients will themselves tell other people (also sworn to secrecy), who will tell still others. Ultimately, the leaks to the press may not come from an insider, but from an insider's confidante, or the confidante's confidante.

Certain well-known processes are likely to distort the story as it spreads. *Leveling* refers to the loss of detail, sometimes significant detail. Qualifications, reservations, and other moderating information may be lost. For example, someone may hear that contact has been made with an aggressive culture that openly vows to eliminate all potential opposition. This may draw attention from an important qualification, such as the realization that their society is safely isolated in an opposite corner of the universe.

Sharpening refers to the accentuation or exaggeration of certain details. When we hear an interesting story, these details stand out and gain further enhancement with the retelling. For example, a civilization that is wealthy or old may become fabulously wealthy or ancient as a result of many retellings of the story.

Assimilation refers to distortions we make to achieve consistency with what we already know. In a sense, our understanding of the world and its workings and the knowledge that we have accumulated over the years give us a template. In the process of assimilation, we try to make new information fit that template, even if this involves twisting or fabricating details. Although we end up with a coherent package of ideas, what we have "learned" is inaccurate. For example, previous reports of UFOs may lead to the description of a UFO as "saucerlike," even though someone viewing it without preconceptions might describe another geometric form. If there are several leaks, there may be several versions that become increasingly dissimilar as they move from person to person or group to group. Thoughtful officials who remain silent while trying to develop an accurate and comprehensive report may be accused of a cover-up.

We cannot predict whether the discovery of extraterrestrial intelligence will remain confidential pending verification of the evidence and careful scientific and political analyses of how to proceed, but we have to acknowledge a possible breach of protocol. This would thwart the goal of an orderly and accurate dissemination of the news.

Official Announcements

The proposed International (SETI) Protocol calls for prompt dissemination of the news to scientists and other people throughout the world. The first person that the searchers are instructed to inform is the secretary-general of the U.N. Next they are supposed to alert the officers of selected scientific societies, who will in turn inform their constituencies. As of yet, the International Protocol has not been accepted by the United Nations and does not have the force of a treaty or law.

The person who makes the announcement to the public will make a tremendous difference in how it is received. Consider, for example, identical reports of unidentified flying objects, one coming from a Nobel Prize–winning physicist, the other from a publicity seeker who has aspirations of becoming a movie star. The speaker's lack of expertise is a common excuse for dismissing UFO reports, even though some of these reports have come from naval officers, pilots, and others with strong credentials.

Two qualities of effective communicators are expertise and trustworthiness.[15] We do not accept what we hear unless we believe that the speakers know what they are talking about and that they are telling the truth. A government agency should have little trouble finding spokespersons who know what they are talking about, but they may have trouble convincing the audience that they are telling the whole and unvarnished truth.

Over the past 40 years or so people's trust in the U.S. government has steadily declined. The 1950s and 1960s saw the Korean War and the Vietnam conflict, which were bewildering to many and seemed to have disappointing outcomes. Senator Joseph McCarthy and others hammered away at the integrity of the government, race riots destroyed parts of several cities, and governmental agencies such as the CIA were involved in embarrassing incidents, including the Bay of Pigs invasion of Cuba. The Watergate scandal of the Nixon years and Irangate of the Reagan years were other pivotal events. The U.S. government has never enjoyed the full faith and confidence of all of its citizens, but Gallup polls conducted between 1958 and 1994 showed a precipitous decline. The percentage of people who reported that they trusted the government in Washington to do what was right "all or most of the time" dropped from 73 percent to 19 percent. It is no longer unthinkable that major institutions could act in ways that seem contrary to the common good.[16]

There are already doubts about the government's credibility in space-related matters. Some people believe that men have never walked upon the moon and that the lunar landings were faked by movie production companies. "Cover-up" is already a blanket explanation for the lack of corroborating evidence for UFO sightings, and as we shall see, whether or not there is a cover-up, some government actions have fed conspiracy theories (see chapter 10).

Confronted with a communication of dubious veracity, people tend to ask themselves, What's in it for the speaker? What is there to gain from presenting this information in this way at the present time? Does this enhance the speaker's wealth, status, or power? If it does, the speaker's truthfulness is questioned. If, however, someone is releasing information that is not in his or her own interest, then the news may be accepted at face value. If I say that I have formed an alliance with friendly space aliens, I may be trying to make myself look good. If, on the other hand, I describe the aliens as intent on plundering our resources, the "look good" alternative no longer holds and I become more believable. Ironically, then, a spokesperson who releases bad news may be seen as more truthful than a spokesperson who releases good news.[17]

Should spokespersons attempt to play down the potentially frightening aspects of a discovery? Or should they instead focus on the kinds of things that could go wrong? The answer to both questions is no. If official communiqués ignore potential threats, the public will not take the situation seriously and may be overwhelmed when the truth is revealed. In the latter case, people's psychological defense mechanisms will swing into play, and they may convince themselves that the message is false or somehow rationalize away the unpleasant reality. In general, moderate threats attract people's attention but do not overwhelm them. An acknowledgment of potential threats, coupled with honest reassurances and suggestions on how to take protective actions, is a promising approach.[18]

Media Accounts

Dorothy Nelkin describes a positive symbiotic relationship between scientists and the media. Scientists help reporters deliver the news and reporters help generate public support and funding for science. The relationship is not flawless: parties on each side occasionally make mistakes, and there are some concessions to self-interest. Scientists sometimes exaggerate the significance of their personal contributions, and as we shall see, news accounts can be premature, incomplete, distorted, or extreme. Yet, while news reporters can and have helped "kill" important science projects, more often than not the media "sell" science.[19]

A 1989 survey by Donald Tarter revealed that media representatives share SETI scientists' views of the importance of SETI and that in their dealings with SETI searchers media representatives tend to be eager and trusting.[20] Both the scientists and the media representatives consider SETI important for stimulating the public's interest in space, and both groups endorse safeguards against premature announcements and misinformation. However, in comparison with the media representatives, the SETI scientists believe that the public holds SETI in higher regard, and they are more optimistic about public response to an announcement.

Many individual reporters maintain high standards and recognize the value of getting the story right. Kendrick Frazier, a reporter, once wrote that "the accuracy and tone of the initial coverage will determine to a large degree the way that the people of the world will react during the first few days and weeks ... [and] shape opinion for years to come."[21] Because sensationalistic reports would give rise to confusion and hysteria, Frazier made a strong plea for comprehensive news gathering and accurate, restrained reporting. But of course, different journalists, broadcasters, and publishers have different standards. Many of the newspapers available at supermarket checkout stands spread disinformation now, and there is little chance that they will mend their ways if ET is actually discovered! Premature release of the story is likely. In Tarter's survey, about half of the media respondents believed that the news story would be too important to hold for scientific verification.[22]

Ian I. Mitroff and Warren Bennis suggest that many news accounts are part fiction.[23] One reason is that major news events are complicated and do not lend themselves to brief presentation. Instead, they require careful study and explanation. Reporters sometimes may lack the knowledge to provide an accurate analysis of a situation, or time pressures (such as publication deadlines or a need to scoop the competition) prevent them from doing so. Reporters, like other people, try to achieve closure or "fill in the gaps" in their knowledge. In the process they turn an incomplete story into something that seems plausible and has a beginning and an end. Reporters who do develop accurate and in-depth news stories are likely to go unrewarded for their efforts because they operate outside the grasp of the mass audience or do not finish their report when it is still newsworthy. Problem number one, then, is that we can expect the media to "keep it short and simple" at the expense of their reports' accuracy and the public's understanding. A related problem is television's desire to seek "instant closure," to finish a story before viewers retire for the night. Thus there are attempts to draw conclusions even before events unfold.

Then there is the mass audience's demand (or perceived demand) for entertainment. A story can have important implications, but it is unlikely to command attention unless it has high interest value. This means capitalizing on its "grabby" or sensationalistic aspects and replacing thoughtful commentary with sound bites and other "hooks" that are emotionally engaging though misleading. Thus, in addition to keeping it short and simple, the media are required to "keep it fun or at least interesting," again at the expense of accuracy and understanding. In the words of Terrence O'Flaherty, "Take a look at the 11 P.M. newscasts and you will see a living continuation of the entertainment programs which have preceded them since 7:30."[24]

Yet another problem is the blurring of reality and fantasy. Technology makes it possible to present things as one would like them to be, rather than as they are. Scenes are reenacted for dramatic effect, pictures (including motion pictures) are retouched or doctored, misleading charts and diagrams are prepared, all in the alleged interest of revealing the truth. In the retouching, the reenactment, the dramatization, fictional elements gain ascendance over factual elements. The audience, conditioned by tabloids and "docu-dramas," doesn't know where fact ends and fiction begins.

We may add yet another complicating factor to Mitroff and Bennis's list. Many organizations and groups will have a vested interest in whether or not intelligent extraterrestrials exist, and if they do exist, the nature and significance of their relationship to us. These vested interests will include religious and political groups, government agencies, even private businesses and hobbyists. Each will be at work assembling the "facts," developing interpretations, and making plans. Each will try to bring their ideas to the attention of the public.

Discovery of intelligent extraterrestrials, the biggest science story of all time, may not have much sticking power in the news. It could rapidly fade from the front pages if additional information is hard to come by, for example, because of delays in translating their transmission. Then the story will be driven from the front pages by faster-paced thrillers with more immediate relevance to everyday life and with greater "human interest." Jerry Kroth pointed out that people attend to news that caters to their narcissism, that is, a "kind of self-centered appreciation of the world in which local and provincial peccadilloes" outweigh significant global events "to the point of absurdity."[25] Kroth points out that many more stories were written about millionaire heiress and kidnap victim Patty Hearst than about the first man on the moon, Neil Armstrong. In 1995 there was more coverage of the O. J. Simpson trial than of a major pestilence in Africa or the war in Bosnia. Certainly, a real-life ET would capture media attention, but faint radio waves bearing arcane messages might not get equal attention.

FORMING IMPRESSIONS

Once contact is acknowledged, psychological processes that affect how people form impressions of one another will also affect our impressions of intelligent extraterrestrials. A near certainty is that not everyone will have the same impressions of the aliens. Another near certainty is that some people, when looking at the diverse attitudes and ways of other people, will be open-minded and tolerant. Others will have great difficulty getting comfortable with totally different life-forms and cultures. Still another near certainty is that people will be influenced by their preconceptions, which will not be based, necessarily, on facts.

Vivian Sobchack suggests that we develop "image repertoires," that is, stores of images and scenarios that help us imagine or interpret situations that we have not actually experienced.[26] The images and scenarios in these repertoires come from myths (for example, the UFO myth, as discussed in chapter 3), literature (for example, science fiction books), and the media (for example, movies such as *ET*, *Alien*, and *Alien Autopsy* and television shows such as *The X-Files* and *Unsolved Mysteries*). Images may bear little or no relationship to the facts, but they still provide an important framework for interpreting new images.

In the United States, Sobchack notes, the mass media have given rise to three images of extraterrestrials. One is the menacing and dominating "colonizing alien" presented in *The Thing*, *The War of the Worlds*, and *Alien* (and might include the alien abductors described in chapter 3). These predators are intent on taking over our territories and inflicting psychological and bodily harm. A second image is that of the benevolent alien representing some higher force that has come to protect our planet or save us from ourselves. These are friendly creatures such as depicted in *Close Encounters of the Third Kind* and *ET* (and may include the benevolent "space brothers" described in chapter 3). A third image is that of the cyborg: part living being and part machine, it has both menacing and protective elements. This image builds on popular conceptions about microchips, biotechnology, and artificial intelligence.

Image repertoires vary from culture to culture, from subculture to subculture, and from person to person. For example, among highly educated people images may come from "highbrow" science fiction writing, whereas among less educated people scenes from comic books and "low-brow" action films may predominate. Image repertoires will color first impressions. For example, a person who has seen the cuddly creatures in *ET* (but not the monsters in *Alien*) may have a more positive first impression than a person who has seen the monsters in *Alien* (but not the cuddly creatures in *ET*). Image repertoires may also counteract rational explanations of events. For example, it may be difficult for scientists to convince people that extraterrestrials 1,000 light-years away are harmless because they can't get from there to here. From reading science fiction people "know" about warp drive, hyperspeeds, time travel, astral projection and other means that "they" might use to end up on our doorstep.

People's first impressions also will to be affected by pragmatic considerations. That is, attitudes toward ET will be determined in part by what people expect to gain or lose as a result of the contact. This follows from the discussion of the effect of rewards and punishments on attitudes as set forth in chapter 1. Certainly philosophers and scientists who have long maintained that there is "life out there" will have been proved right, and this may promote positive impressions on their part. Philosophers and scientists who have adopted the "we are alone" stance will have been proved wrong, and this unpalatable fact might contribute to unfavorable views on their part. Similarly, we might expect a positive reaction from people who believe they have new opportunities as the result of the contact. Those who have marketable skills that can be bought to good use in dealing with ET are likely to have views positively colored by this fact. We can expect positive views on the part of those who believe they can interpret alien messages, or who see some sort of opportunity to engage in commerce with the aliens. People who stand to lose because alien technology threatens to put them out of work would be expected to react less favorably.

David Sears and Rick Kosterman have recently reviewed the effects of the mass media on shaping public opinion, and their conclusions may be helpful for understanding people's first impressions of extraterrestrials.[27] In the 1930s, social scientists tended to view the media as having profound and pervasive effects on people. Audiences of that day were thought to be passive and impressionable; reactions to the *War of the Worlds* broadcast and the apparent ability of dictators to sway mass audiences betokened the media's immense power. This early model suggested that media accounts would effectively determine our initial impressions, and because we are all exposed to the same media, we would be likely to form similar first impressions.

By the mid-1940s, however, opinion researchers decided that they had overestimated the media's persuasiveness. Research in the 1940s and 1950s showed that people tended not to listen to broadcasts that would challenge their views. When they were exposed to information they didn't like, they tended not to notice it; if they noticed it, they didn't remember it; if they remembered it, they did not change their minds. Over the years other findings supported this "minimal effects" model. For example, although television sets may be on for many hours a day, members of the "viewing audience" often engage in distracting activities (talking, eating, reading, and so on). In some homes, televisions run hour after hour with nobody even in the same room. The expectation in the 1950s was that media accounts would not have much of an effect.

Today, note Sears and Kosterman, we conclude that the mass media have minimal effects on people's attitudes *except* when the person doesn't feel strongly one way or the other *and* there is a concentrated media blitz to push them in a particular direction. If the media mount an intensive campaign to present the extraterrestrials in a favorable or unfavorable light, it is likely to have a strong impact, but only on people who haven't given the matter much thought. As for people who already have an opinion, we will do well to remember Mary Connors' hypothesis that preexisting beliefs will be the best predictors of people's immediate postcontact reactions.[28]

Physical Appearance

If we don't know what they look like, we will try to find out. Physical appearances have a powerful effect in many contexts, and it is likely that a species that is ugly by our standards would have a disadvantage. Science fiction writers and movie producers are well aware of how appearances affect us, and they are able to construct their aliens to achieve the desired dramatic effects. We may respond favorably to aliens that remind us of children or pets, and less favorably to those that remind us of lizards, slugs, octopuses, or other creatures that have "image" problems.

Reactions to certain appearances are genetically wired into a species. For many animals, large eyes or eyelike patterns tend to elicit fear and escape responses. One possible explanation for this is that large, staring eyes often belong to predators. Immense eyes or eyelike appendages may also be threatening to humans, and this threat may contribute to an aversion to "bug-eyed monsters."

Also threatening are long, sharp teeth (suitable for carnivorous dining) and certain postures and gestures associated with attack. A hardwired favorable reaction is our response to a "Kewpie doll," with its relatively large head, large eyes, tiny button nose, small chin, and small puckered mouth. This facial configuration, which we associate with infants, tends to trigger maternal and protective behaviors.

Perhaps some of our features will elicit preprogrammed reactions from them. As erect bipeds, we could be relatively tall and loom over members of a short, squat species; both large size and looming behavior are frightening to many terrestrial animals. If the aliens lack bilateral or radial symmetry, the redundancy in our system could be unnerving. From the perspective of a cyclops, our pair of eyes would not only give us an advantage in depth perception but also allow us to continue fighting despite an injury to one eye. Similarly, *Homo sapiens*, with its two arms, could seem formidable to a creature that had only one limb. As aggressors, we could use one of our arms to immobilize their single limb while our other arm went on the attack. It doesn't entirely matter that neither they nor we may have any intention of pressing an advantage, or that given modern technology (including weapons), big eyes or sharp fangs are not much of an advantage at all.

Although we tend to like people who are similar to ourselves, a hominoid appearance would not necessarily work to their advantage. For over a century the image of the "hairy ape man" has been used to ridicule various racial, ethnic, and national groups, and to suggest weak moral fiber and low intelligence. This image of a sometimes weak, sometimes strong simian figure with a sloping forehead and prominent jaw has become, according to Joseph Bulgatz, the standard characterization of the enemy in time of war. Before 1900 this image—which suggests that the enemy is lower down on the evolutionary ladder—was used by both the Irish and the English to lambaste each other. It was elevated to a high art form in American caricatures of the Japanese during World War II, and was invoked once again in cartoonist David Levine's caricature of Saddam Hussein that appeared in the *New York Times* during Operation Desert Storm.[29] This is the image also of "bigfoot" in the Pacific Northwest and the "abominable snow man" in the Himalayas. The point is that as a result of cultural conditioning we have powerful image repertoires stored within our heads and the extraterrestrials' physical appearance could evoke these images, whether or not they should apply.

Salience of Negative Information

When we form impressions, we tend to assign great weight to other people's unflattering or otherwise undesirable characteristics. Their real or imagined shortcomings will play pivotal roles in our evaluations of them. Negative information is usually considered more authentic and informative than positive information. When we combine different bits and pieces of information to form an overall picture we give the negative information the greatest weight.[30] Our tendencies to "accentuate the negative" may be hardwired into our species. By causing us to be suspicious and cautious in the face of potential danger, it may confer a survival advantage. The downside is that it sometimes leads to unnecessarily harsh evaluations of other folk.

Human reactions to extraterrestrials' seemingly human weaknesses could play a major part in our overall impressions. The ancient Greeks, noted sociologist Erving Goffman, cut or burnt signs into living human bodies to advertise that the bearer was a slave, criminal, or traitor—a person to be avoided, especially in public places. In contemporary times many people respond adversely to stigmatized individuals.[31] What if alien evolution or culture confers attributes that, in contemporary human society, are stigmatizing? Such characteristics could capture our attention, control our overall impression (other characteristics would recede into insignificance), and pose a formidable barrier to friendly relations. Let us consider a few possibilities.

Perhaps extraterrestrials could evolve in such a way that they appear physically handicapped to human observers. For example, a planet with low gravity, or a highly technological culture where manual activity was no longer required, could give rise to creatures with extremely thin or even atrophied limbs. Such creatures might find it difficult to move about on Earth, perhaps forced to rely on some sort of prosthetic device such as a wheelchair. Then, too, aliens may be "blind" on Earth because their "eyes" are geared for wavelengths outside ours, or their "ears" may be tuned to frequencies that are below or above the thresholds of normal human hearing. All of these characteristics might be quite adaptive on the home planet, but would seem to reflect abnormalities or disabilities here.

Given the assumption of our mediocrity, it seems implausible that any creature we hear from could be dim-witted or slow. However, this perception could come about in any number of ways. As Burghardt has pointed out, cold-blooded animals such as reptiles are less rapid processors of information than mammals are.[32] In general, sluggishness is considered indicative of inferior intelligence, but it might only mean that their bodies are not tuned for our terrestrial conditions or that they operate on a different internal clock.

It is also conceivable that extraterrestrials will display behavior that is normal and adaptive given their ecological niche, but that is suggestive of mental illness to us. Perhaps ETs are less or more emotional than most humans, and their humor may seem inappropriate. An interesting theme in alleged alien–human contact is the unusual behavior of the alien, as if it had a mental screw loose or that despite its humanoid appearance it was somehow out of touch. Of course, this can happen to any stranger who lacks familiarity with local laws, traditions, and customs, but then again, not all people are fond of strangers, either.

A common theme in both science fiction and discussions of alien–human contact is that the aliens maintain political or religious views that fall outside the range deemed acceptable by many humans. For example, incest is widely condemned on Earth, yet an alien culture might tolerate or even encourage intrafamilial marriage. (Some experts believe that as few as ten people would provide an acceptable breeding population for a founding colony such as a worldship.[33]) Routine cannibalism, abortion, infanticide, patricide, and the like could serve to define aliens as unacceptable in some people's minds, as could an adherence to religions that require the sacrifice of living beings.

Many of us believe that we live in a fair and equitable world.[34] If ET appears to suffer from some sort of limitation or affliction, observers must decide either that the universe is cruel or that the afflicted parties somehow deserve their fate. Those of us who are motivated to maintain a belief in a just world sometimes devalue or derogate the victim. In other words, we must guard against a tendency to convince ourselves that members of a species stigmatized by human standards "got what they deserved."

Social Categorization

People have a tendency to classify one another into categories, and as this occurs, two groups develop. These are the in-group (one's own group or kind) and the out-group (the other's group or kind).[35] Once categorization occurs, a bias develops for the in-group at the expense of the out-group. Although complexity and social diversity are recognized within one's in-group, the out-group tends to be seen as homogeneous; that is, "they are all alike" (even though they may be as different as a Saint Bernard and a poodle). Furthermore, knowing that another being belongs to a particular category, we tend to make sweeping generalizations as to what he, she, or it is like. We tend to see members of the out-group as unintelligent, immoral, and blameworthy. The categorization process could lead to simplistic thinking about extraterrestrials, a failure to recognize differences among them, and a tendency to devalue them relative to ourselves.

Social categorization plays a role in prejudicial attitudes in that we evaluate and treat other people on the basis of their membership in a social category rather than on the basis of their individual qualities. Prejudicial attitudes are negative attitudes that depart from one or more of three ideal norms: the norm of rationality, the norm of justice, and the norm of human-heartedness.[36] The norm of rationality suggests that we should be accurate and factually correct, logical in our reasoning, and cautious when making judgments. A prejudiced attitude is often inaccurate, incorrect, and illogical. The norm of justice suggests that all people should be treated equally. A prejudiced attitude includes the belief that differential treatment should be based on group membership rather than on individual ability. The norm of human-heartedness prescribes tolerance and compassion. A prejudiced attitude often advocates kicking, rather than rooting for, the underdog.

Prejudice on Earth is complex and derives from many factors, economic and historical as well as social and psychological. Perhaps, when contact is made, we will start with a clean slate. We may think that we are ready to accept aliens with green, leathery skin or bright orange skin with purple splotches, or no skin at all. Yet, how many people who believe they could get along with such creatures would be equally receptive to extraterrestrials whose skin colors happened to be one of those colors (white, black, brown, red, or yellow) that are found among peoples on Earth but that differ from their own?

Political and Historical Contexts

Politics and history shape expectations about aliens.[37] A good example is the fear and paranoia of the McCarthy era of the early 1950s. At that time there was a sense that a Communist lurked behind every rock, ready to overthrow the government of the United States. The fictional aliens of that time tended to resemble the American and western Europeans' views of Communists. Aliens were portrayed as powerful, evil, and intent on a world takeover. Representative here are the podlike creatures in *Invasion of the Body Snatchers*, who took over a human body as soon as its natural occupant (so to speak) fell asleep. Later, in the relaxed political climate of the 1970s and 1980s, we find kinder creatures such as ET.

The hypothesis that decreased world tensions are associated with more positive views of aliens was dealt a severe blow in 1996. *Independence Day*, a science fiction film about the invasion of Earth by predatory aliens, drew record audiences. This film and others such as *The Arrival* and *Species*, and a large roster of new television programs, built on people's fear. Nevertheless, a period of intense brinksmanship or conflict at the time of detection could affect our impressions of ET. On the one hand, ET might be seen as yet another threat. On the other hand, ET might be seen as a powerful and just ally poised to come to the rescue. The latter view might be particularly likely in the weaker camp if desperation encourages a search for magical solutions. The "winning" side could see the aliens as strong and just allies, or as introducing a joker into the deck.

The basic rule of thumb is that all, other factors being equal, human reactions will be more positive in good times than in bad. The sheer association of extraterrestrials with peace and prosperity (even in the absence of a causal relationship) is likely to have a positive effect. That is, the good mood associated with general affluence is likely to rub off on our perception of ET. Of course, the process can work the other way. If extraterrestrials are associated with war or economic depression, attitudes toward them will be less positive.

In difficult times, people tend to take out their frustrations on other people or even on impersonal entities. Harsh economic periods are associated with high degrees of prejudice, rejection of the out-group, and direct aggression. For example, at one time in the American South, the declining price of cotton was associated with increased violence, including lynchings, toward blacks. Some frustrations are taken out on nonhuman entities. After Washington's Mount Saint Helens erupted, it became a scapegoat even for unrelated problems. Nearby residents blamed the eruption for "bad moods, illness, disagreements, and even drunkenness."[38]

Social Factors

Finally, the reactions of other people will have a major effect on our reactions to ETs. There is a strong tendency to mimic other people's emotional reactions; thus, after the first person panics or proclaims salvation, there will be a tendency for other people to do the same. There are also pressures to conform.[39] To secure social approval and avoid rejection, then, many people will adopt their friends' attitudes. Most important, when we are confronted with uncertainty, when we do not know how to feel or react, we look to others for cues that help us define our emotional experiences. We are more likely to do this when the situation is unusual, unfamiliar, uncertain, or ambiguous. The less "hard" information that is available, the greater the role of other people's reactions.[40] Because under many contact scenarios we can expect scant hard information, people are likely to turn to one another for ideas to clarify the situation and for help in putting their feelings in perspective.

CONCLUSION

Since we don't know what "they" will be like, it is difficult to forecast the impressions that "we" will form. Yet we can identify some of the processes that are likely to swing into play. Most likely, we will begin with a radio intercept that is skimpy or only dimly understood. The facts, such as they are, will be filtered and augmented by one or more rumor mills and by the media. Our impressions will reflect our experiences and expectations, the images and prejudices that we carry around in our heads, other people's perspectives, and political-historical contexts.

Will everyone be influenced by superficialities such as appearance? Extrapolating from a recent review by Richard Petty and his associates, we might expect that some people—those who are somewhat cerebral or intellectual to begin with, who feel some responsibility for making informed decisions, and who see for themselves the implications of the discovery—are likely to look beyond superficialities, await additional information, and think it over before drawing any conclusions.[41] Once these people form an impression, however, it will be detailed, highly integrated with their other views, and resistant to change.

Other people, who are more impulsive or reflexive and who do not believe that they have any personal stake in the unfolding events may be susceptible to physical appearance, media commentaries, their friends' evaluations, and other cues that help define ET's nature. Although their intellectual style and lack of personal involvement cause them to form an immediate opinion, this opinion will not be well integrated into their overall worldview and is relatively susceptible to change.

Although an unthinking approach to forming an attitude seems stupid or lazy, we do not always have the time and energy to assemble and weigh information and decide what is good and what is bad. Instead, we must conserve our intellectual resources by relying on cues, sometimes superficial cues, for dealing with situations that confront us. This laziness may characterize the processes that will form some people's initial impressions of extraterrestrials and their civilizations.

While the lazy approach to developing attitudes saves time and energy and keeps people from being entirely overwhelmed with information, it can lead to inadequate and occasionally outright misleading pictures of science, politics, and international events. Simplistic explanations, limited attention spans, a desire to be entertained, and the technical means to tamper with the evidence encourage simple-minded thinking, misperceptions and misunderstandings. There is no reason to believe that such an approach will disappear when the news is relayed of contact with an extraterrestrial civilization. People who lack the ability or motivation to think it through will base their views on such criteria as the credibility of different spokespersons, the number (not the quality) of the arguments, and social consensus.

NINE

INITIAL IMPACT

Upon learning that Martians had invaded New Jersey, a young man attempted to call his girlfriend in Poughkeepsie, but found this impossible because the telephone lines were jammed. He and his roommate had heard that Princeton, his parents' hometown, had been obliterated, so there wasn't anything they could do there. They thus set off for Poughkeepsie, the young man flooring the accelerator and the roommate crying and praying. They averaged over 70 miles per hour, didn't notice passing through certain major towns, and later wondered why they hadn't been killed. In New York, a woman who had "never hugged her radio so closely" as she did that night clutched a crucifix and prayed. She left her window open so that she could sense the first whiff of poison gas, whereupon she would try to seal the room with waterproof cement until the gas dissipated. When she heard that the alien invaders were crossing the Hudson River in her direction, she was torn between "running up to the roof to see what they looked like" and staying by the radio to keep abreast of their whereabouts.[1]

These accounts are among those collected by Hadley Cantril in his classic study of people's reactions to Orson Welles's *War of the Worlds* broadcast in 1938. As the broadcast continued, many listeners hit the road; according to legend, some did not reappear for days. Because it deals with an invasion from Mars, it may appear that Welles's radio play brought about an excellent demonstration of people's likely reaction to contact. In fact, that scenario differs so substantially from the SETI scenario that it may have almost no predictive value at all.

Mary Connors draws a useful distinction between the short-term and long-term effects of the discovery of extraterrestrial intelligence.[2] If contact occurs, short-term effects will begin as soon as word of the discovery is accepted. These will be evident in people's emotional states, an inundation of media reports and publications, clogged switchboards, and intense pressures on research institutions and governmental agencies. The immediate impact of contact will be the acute phase of our response. It will rapidly spread and peak as the news spreads across the globe, but will partially subside if the news stalls due to a slow pace of developments following the initial announcement. The short-term impact will make itself known within minutes or hours of the announcement of the discovery and probably begin to subside within weeks or even days.

The long-term impact may begin before the initial excitement wanes, but it will follow a slow upward trajectory and could take decades or years to reach full stride. Over the long haul, contact with ET could prompt fundamental and far-reaching consequences for individuals, societies, and supranational systems. Over the years, contact could affect our arts, literature, science, technology, philosophy, and religion—in short, every aspect of our culture. In this chapter we will focus on the short-term, or acute, phase of human reactions to contact; in the next chapters, on consequences that could evolve over long spans of time.

One way of organizing possible responses is along a continuum, with positive emotional reactions at one end and negative emotional responses at the other.[3] At the positive extreme we find hope and euphoria. This can occur if the discovery confirms preexisting optimistic beliefs or seems, in some way, to offer prospects for a better world. We know that people are dominated, sometimes, by wishful thinking. There have always been people who want magical solutions to problems, and perhaps ETs will join the ranks of parents, the gods, and Lady Luck. Certainly, the idea of extraterrestrials as father-figures or "gods from space" is the theme of many science fiction stories and a presupposition of many religious cults.

At the midpoint of the scale is indifference. Especially under a SETI-like scenario, which may involve little more than the detection of a carrier wave, scientific chatter, or a courtesy "Hello," the discovery could have few immediate implications for people's lives. Some of us are not all that interested in history in the making, or are so caught up in the minutiae of everyday survival that "the greatest discovery of all time" won't really matter.

At the negative end we find apprehension, fear, perhaps terror. People's image repertoires contain many images of grotesque predators, and a common theme in science fiction is that if the ETs don't start trouble, then human authorities do. Fear is of great concern because, if widespread, it could lead to panic and rupture the fabric of society.

In fact, we might expect any or all of these reactions, depending on the nature of the extraterrestrials, the people involved, and the unfolding of the contact scenario. Our reactions will depend on such factors as our cultures and our personalities, our perceptions of the aliens, whether they seem to augur well or bode ill for humans, and, in the latter case, our confidence in our ability to counter the threat.

PRECEDENTS

Although somewhat crude and imperfect, a few historical precedents shed some light on human reaction to contact. These incidents involved significant numbers of people who believed that humans had detected (or had been detected by) intelligent extraterrestrials. These episodes differed greatly from one another, and so did people's reactions.

Bat Men on the Moon (1835)

In the 1830s, many respectable scientists believed that there was life on the moon, and some had claimed that they had seen roads and other artificial features on the moon's surface. As recounted recently by Joseph Bulgatz, this set the stage for a major hoax.[4] On August 25, 1835, the *New York Sun* began a series of articles purportedly based on those in the *Edinburgh Courant* and the "highly respected" *Edinburgh Journal of Science*. These sources stated (the *Sun* reported) that the well-known discoverer of Uranus, Sir John Herschel, had developed a telescope with a 7-ton objective lens. Through an ingenious arrangement of arc lights, lenses, and mirrors, Sir John was able to magnify an image 42,000 times with no loss in brilliance.

In six clever installments the *Sun* wove a fascinating story of Sir John's observations of the moon. He had discovered a fairy-tale setting with active volcanoes, blue inland seas with wide white beaches, vast green plains, trees similar to yews and firs, dark red flowers, green marble, and amethyst obelisks. In this setting he saw a diversity of exciting life-forms. The animals and birds were reminiscent of real and mythical animals, but they differed in significant ways (for instance, the "beaver" was a biped) and generally had a spectacular appearance. Among the more intriguing animals were a goat-size creature with a unicorn horn and another "strange amphibious creature of a spherical form which rolled with great velocity across the pebbly beach."[5]

Then appeared "*Vespertilio-homo*," or "bat man." These creatures, about 4 feet tall (the same height as today's "alien abductors") flew as bats but walked as bipeds. The *Sun* described them as intelligent, expressive and communicative, and capable of fine arts and architecture. Larger versions of this creature, clustered around a beautiful temple made of blue stone with a golden-hued roof, in a scene "evocative of paradise," these "happy, angel-like creatures" happily shared food and "engaged in innocent activities such as eating, flying, bathing, and lolling about, while other animals wandered among them without fear."[6]

The public was hooked, and by the fourth installment not only had the *Sun* won the New York newspaper circulation contest, but its readership had also surpassed that of the *London Times*. There was in fact a famous astronomer by the name of Sir John Herschel, and the Edinburgh sources were real, if not accurately cited. The real author of the fiction was one Richard Adams Locke, who was astounded to overhear strangers verify and support various aspects of his made-up story. The *Sun* articles were reprinted and widely read, and they appeared, off and on, around the world, for about a 50-year period, even though it was only a matter of days before the *New York Herald* blew the whistle on the *Sun*'s hoax. An admiring fellow hoaxer, Edgar Allan Poe, estimated that nine out of ten people fell for Locke's story, even though Poe himself found many implausibilities and contradictions within it.

The readers were delighted with the discovery of life on the moon and excited by the richly detailed descriptions. The bat men on the moon were attractive and peaceful and, in any case, unaware of our existence and safely sequestered far from Earth. There was no threat whatsoever, so the path from detection to panic was never taken.

Canals of Mars (Circa 1895)

Through perception we organize and interpret the evidence that confronts our senses. The resulting impressions reflect not only the objective properties of what we observe but also the properties of the situation and of the observer. Scientific procedures provide reality checks, but scientists are not immune from the principle that what we see is, to some extent, in the eyes of the beholder.

William Sheehan has shown how the actual or "objective" characteristics of planets combine with the size and quality of the telescope, with atmospheric and other viewing conditions, and with the astronomer's eyesight and expectations to shape the perception of planets.[7] In the late 1800s, the work of three influential scientists—Giovanni Schiaparelli of Italy, Camille Flammarion of France and Percival Lowell of the United States—fostered a belief in advanced life-forms on Mars.

By the 1870s, astronomers had glimpsed mysterious lines on Mars, and the term *canale* had been introduced to describe them. In the latter part of that decade, Schiaparelli had a pronounced tendency to perceive Martian features as geometric forms. On September 15, 1877, he identified his first "canal" and over the ensuing years identified many more. Although the canals were somewhat hard to discern, over time Schiaparelli's drawings showed them as straighter and more pronounced. Initially, "canal" was a simple descriptive term, but by 1893 Schiaparelli decided that the Martian canals were in fact a water-distribution system. Yet he was not willing to conclude that this system was the work of intelligent beings.

Schiaparelli's observations influenced Camille Flammarion, whose interest in astronomy was piqued at the age of 5, when he observed an eclipse. By 16 Flammarion was a junior assistant to Urbain-Jean-Joseph Le Verrier at the Paris Observatory. Although for the rest of his life Flammarion viewed himself as an astronomer first, he was fired before his 20th birthday for publishing a book claiming that planets were inhabited. Later, in *La Planète Mars*, he argued that the red planet was very much alive and that the canals were indeed waterways. Flammarion was a phenomenal success; one wealthy admirer gave him a chateau and a large telescope.

Flammarion believed that where life can arise it does arise, that life-forms on different planets achieve different levels of development, that *Homo sapiens* is far from the highest rung on the cosmic evolutionary ladder, and that we should rejoice at the abundance of life throughout the universe.[8]

Flammarion's chateau became "a nerve center for Martian research."[9] A wealthy, brilliant Harvard graduate, Percival Lowell, visited there regularly, where he addressed Flammarion as "my Martian friend." In Arizona, Lowell established an observatory under some of the clearest sky in the country. In 1894 Lowell saw *his* first canal (or at least green vegetation growing near the canal), and over the years he and his associates found much to support the conclusion that there was intelligent life on Mars. Always at the heart of the matter were the canals: to him, their absolute straightness, uniform width, and systematic radiation from the oases proved conclusively that they could not be random.[10] Lowell was convinced that as Mars became increasingly dry its inhabitants built the canal system to distribute its last supplies of water.

Through a series of books, articles, and lecture tours, Lowell publicized his vision of life on Mars: The intricate system of canals, locks, and dikes was built from scratch by an advanced race whose physical strength and mental powers eclipsed ours. Although they were ruled by an elite, they were friendly and peace loving. Most scientists dismissed Lowell's conclusions, but his lectures were jam-packed and public reaction was favorable, ranging, according to one account, from "accepting wonder to skeptical but tolerant amusement."[11]

As the idea of life on Mars became popularized, plans emerged for attracting the Martians' attention by building mammoth bonfires; igniting huge, intricate systems of trenches that had been filled with kerosene; planting forests whose unnatural patterns would suggest life on our planet; and, later on, radio transmissions. Lowell's description of Martians safely isolated on a dying planet generated interest and excitement, but did not cause widespread fear or panic. The issue was how to make ourselves known to the Martians, not how to hide from them.

Lowell stood by his conclusion and viewed criticism as the disgruntled reactions of scientists who were clinging desperately to outmoded ideas. Today, observers have difficulty finding the canals of Mars. The smooth and seemingly regular forms of the canals (which seemed to prove that they could not be natural) must have been constructed by the eye and the brain, a human tendency to turn chaos into order and to perceive disordered or incomplete patterns as flowing and continuous. The straightness, symmetry, and pattern existed in the eyes of the beholders. As Sheehan points out, the tremendous leeway that expectations and imagination had in affecting perceptions is evident in reports (including Lowell's) that the thin, faint canals could be seen only at certain times and then only for a "flash"—and that they were impossible to remember.

The War of the Worlds (1938–1994)

As already noted, the most celebrated prototype of human reaction to contact with aliens was the public reaction to Orson Welles's *War of the Worlds*, presented by *The Mercury Theater of the Air* and broadcast to a national audience on Halloween evening in 1938. H. G. Wells's novel of the same name, *The War of the Worlds* described a successful invasion of New Jersey by hostile Martians. It was formatted as an unfolding news story, with reporters presenting eyewitness reports and fast-breaking developments from the field.[12]

Competing with CBS's *Mercury Theater* that Sunday evening was NBC's immensely popular *Chase and Sanborn Hour*, starring ventriloquist Edgar Bergen and his dummy, Charlie McCarthy. Shortly into this show, Nelson Eddy began singing, and even as we flip channels today, the radio listeners of 1938 began twirling the dial. As they tuned in to the *Mercury Theater* they heard a startling announcement. An on-the-scene reporter was describing the arrival of a large metal tube in Grovers Mill, New Jersey. There followed, interspersed with brief musical interludes, more stunning announcements and "live news reports" detailing a series of events: identification of the tubes as spaceships from Mars; the arrival on the scene of a supremely confident New Jersey militia; the total rout of the militia by tall Martian robots; Martian advancement on and the destruction of Manhattan; and the revelation that other groups of Martian invaders had landed elsewhere throughout the country. Although by that time Lowell's beliefs about life on Mars had been discredited and although the radio play's time frame should have been a giveaway (only a few minutes elapsed between the arrival of the tubes and the fall of New York City) some of the 4 million listeners accepted the news at face value. Events were canceled in midstream, some people barricaded themselves in rooms, some ran in all directions, and some went into hiding. More constructive responses came from those who volunteered to fight the invaders, activated relief agencies, and offered to donate food and clothing for the homeless. Most people were never fooled at all, or were reassured when their attempts to verify the information failed to confirm the invasion.

As social psychologist Hadley Cantril pointed out shortly afterward, many factors enhanced the illusion of reality.[13] Listeners dialing from Bergen and McCarthy to the *Mercury Theater* did not hear the disclaimer that opened the show. The musical interludes of "Ramon Raquello" and his orchestra did not arouse suspicion, because technical failures were common and stations that were temporarily off-line played music as a filler until the network connection was restored. Live reporting was innovative at that time; audiences were just getting used to it and were not practiced at distinguishing simulations from actual events. One of the most important factors was the seeming realism of the broadcasts. As one man from the Bronx told the *New York Times*, "When I heard the names and the titles of Federal, State, and municipal officials, and when the Secretary of the Interior was introduced, I was convinced that it was the McCoy. I ran out into the street with scores of others, and found people running in all directions."[14]

The temper of the times was another influential element. Nine years earlier the stock market had crashed, triggering a depression that had only recently begun to lift. Many of the listeners had been unemployed or on short rations for years. Another factor was the gathering war clouds in Europe. Hitler was on the march. In a recent discussion based on Carl Jung's analytical psychology, Jerry Kroth suggests that the *War of the Worlds* broadcast resonated with the audience's fear of Nazi aggression. Kroth's analysis draws many parallels between the Martian and Nazi aggression. The Nazi symbolism in the radio play included Teutonic war gods, storm troopers, iron crosses, swastikas, and anti-Semitism.[15]

As Joseph Bulgatz describes, over the next 50 years *The War of the Worlds* was rebroadcast in original and updated forms, sometimes with comparable or even scarier results.[16] A 1944 broadcast in Chile terrified people, causing serious stress reactions including heart attacks, and triggered riots in Valparaiso. A February 1949 broadcast in Ecuador also generated panic. Once the fear-crazed crowd learned they had been duped, their fear turned to anger and they torched the broadcasting station. (Because fear challenges our sense of self-worth and makes us look bad to ourselves, the same conditions that scare us also make us angry.) Rioters kept firemen from putting out the fire—6 people died and 15 were injured. A 1968 rebroadcast in New York City was uneventful, but a 1974 broadcast in Providence, Rhode Island, caused consternation and generated complaints to the Federal Communications Commission. In 1988, a broadcast by Radio Braga panicked listeners in the northern part of Portugal and, as in Quito, resulted in a crowd storming the offending radio station.

Quasars and Pulsars (1960s)

On April 13, 1965, two articles appeared side by side on the front page of the *New York Times*. One of these announced a news release from the Soviet news service, TASS. This article suggested that the Russian astronomer, Gennady B. Sholomitsky had attributed "flickering" radio waves emitted by the stellar object known to us as CTA 102 to a supercivilization. The flickering was the equivalent of switching on and off the energy output of 10,000 billion suns.[17] The adjacent "twin" article by science writer Walter

Sullivan urged caution and noted that according to another Russian astronomer, the much revered Iosef Shklovskii, the sounds emitted from CTA 102 might have been of "natural origin." The *Times* readers of that day were cautioned by both American and Russian astronomers. Len Carter, of the British Interplanetary Society, pointed out, "All that the Russians are saying is that a regular pulse of radio emissions has been discovered.... There are many possible explanations, some of them in purely physical terms."[18] Thus, this false alarm was nipped in the bud before it had a chance to get started. Today, CTA 102 and its peers are known as quasars, brilliant bluish starlike objects that have approximately 10^9 times the mass of the sun and occupy about the same amount of space as our solar system does. The source of their energy is unknown.[19]

Another false alarm occurred in 1967. A young graduate student from Belfast, Jocelyn Bell, was at the controls of a radio telescope at Cambridge University, collecting data for her dissertation under the direction of Anthony Hewish. She noted some "odd pulsating noises being picked up by the receiver, like the kind that spoil TV programs while turning radio astronomers' hair gray."[20] Of particular interest was the unwavering and precise periodicity of the bursts of energy: exactly one burst every 1.33730133 seconds.[21] These emissions were studied for months, and the hypothesis that they reflected intelligent extraterrestrial life was among the contenders. In fact, this object and three similar pulsating objects were designated LGM1–LGM4 (with "LGM," of course, standing for Little Green Men). The researchers held off publishing their findings for the better part of a year, but when they did, according to Thomas McDonough, "pandemonium reigned in the astronomical world" adding that pandemonium in the astronomical world is very mild by everybody else's standards.[22] McDonough continued: "The public wasn't ignorant of the excitement. At first the world was presented with a rash of press reports from scientists announcing that this looked like the real thing at least: the long-awaited signals from little green men."

Unlike UFOs, notes McDonough, microwave observations can be checked scientifically. In fairly short order scientists concluded that these emissions were of extraterrestrial but not of intelligent origin. Celestial radio sources that produce precise, intense short bursts of radio emissions are now known as pulsars. Since the 1960s over 330 have been discovered, and it is estimated that there must be hundreds of thousands of these in our own galaxy, most too far away to be detected by our radio telescopes.[23]

Despite press coverage of quasars and pulsars, there was no reaction comparable to that following the *War of the Worlds* broadcast. A search of my field's best bibliographic data base, *Psychological Abstracts*, reveals several papers on *The War of the Worlds*, including some published in the 1990s, but nothing whatsoever on human reactions to the false alarms of the 1960s. Certainly there may have been interest and excitement, but there was no panic. Of the people who I know were alive at the time, a decent proportion remember Orson Welles's broadcast. I cannot make the same claim about the memorability of the LGM announcements.

UFOs and Alien Abductions (Present)

As noted in chapter 3, many people today believe that humans have been or are in contact with extraterrestrials. These believers represent both genders, a full spectrum of ages, an array of cultures, and many walks of life. Some of these people believe they themselves have been abducted by aliens and taken to alien spacecraft. In most but not all cases, they believe their memories of the episodes are initially repressed but return later under hypnosis. In some cases the emotional reactions that accompany the recollections are very negative and intense, not unlike those that accompany the recollection of a major life trauma such as combat, torture, a natural disaster, or sexual abuse. In other cases, the emotional reactions are not particularly intense or are even construed as positive or fulfilling. Whereas all of this is troubling to some people, there is no widespread panic.

COPING WITH ET

Research on threat, stress, and coping suggests that many lines of defense will keep us from being overwhelmed psychologically if we discover ET.[24] *Threat* refers to the perception of danger or possible harm. *Stress* refers to very high levels of demand upon an individual as evidenced by physiological measures (such as heart rate, blood pressure, blood chemistry, and pupil dilation) and feeling-states (such as anticipation, anxiety, fear, and euphoria). *Coping* refers to cognitive or intellectual processes and to behaviors that help us manage problems and their attendant emotions. Coping involves assembling information about the situation, maintaining physiological and psychological states that are conducive to problem-solving, and preserving freedom of action. Through coping we master situations, overcome damage, or at least develop a tolerance for living with the aftermath. We have many ways to deal constructively with potentially adverse situations, and it is only after multiple failures that we will panic or become mentally ill.

Sometimes we imagine a sharp dividing line between intellectual and emotional responses. But according to Richard and Bernice Lazarus, intelligence and emotions "go hand in hand, and that is why humans, highly intelligent beings, are such emotional animals."[25] Although on occasion our emotions get us into trouble, "emotions are a vital tool for getting along in the world ... they help us decide ... whether we are in danger, safe, or in a position to capitalize on [a situation].... Emotions are intimately connected with the fate of our struggles to adapt to life in a world that is not very forgiving of failure."[26]

Thus, emotions tell us how we are doing. They allow us to quickly redirect our attention to potential emergencies. They help us focus our attention on problems and they give us the energy to act. Certainly, we tend to admire people who can maintain their cool under stressful circumstances, but in the absence of all emotion such a person would not undertake preventive or corrective action and probably would not survive.

According to the Lazaruses, we are continually scanning the world around us, alert for conditions and events that might affect our welfare. Any environmental event that attracts our attention—including the news of an interstellar broadcast, or the landing of a UFO outside the bedroom window—is evaluated for its implications for oneself. More than a century of psychological research shows that new, unusual, or surprising stimuli gain our attention.[27] Extraterrestrial life-forms will certainly be new and unusual, and they could very easily be surprising, so initially will gain substantial attention.

Appraising Threat

Our reactions are moderated by appraisal, that is, the intellectual process we use to size up situations and their personal relevance. Once something (such as a newspaper headline or a UFOnaut) grabs our attention, we are likely to ask ourselves, "Is this relevant for me?" and, if so, "How?" Are the aliens a source of potential benefits—for example, would they want to be our friends and offer us aid? Or are they intent on exploiting or hurting us?

Our estimates of their threat potential are likely to rest on three things: their apparent technical capabilities, their proximity or immediacy, and our perceptions of their intentions. If we believe that their technology is not much better than ours, that they are safely sequestered in some remote corner of the universe, and that they are basically "good guys," we will feel less threatened than if they have an overwhelming technical superiority, are already in our neighborhood, and seem unfavorably disposed toward humans.

As stressed repeatedly throughout this book, we expect that any ETs that we detect will be technologically advanced relative to ourselves. Although their superior technology may have evolved in a peaceful fashion, many of our own important breakthroughs and inventions have been inspired by war. I refer not only to the long bow, the catapult, and the cannon, but also to the tremendous technological advances of World War II: radar, liquid-fueled ballistic missiles, and atomic weapons, to mention but a few. During the 1980s, some of the defense technology associated with the Strategic Defense Initiative was referred to as "Star Wars." Some of this technology may have appeared anyway, but it is extremely doubtful that it would have appeared at the same rate as in times of tension or war. Whatever the roots of their technology, we expect any civilization with the capability of contacting us to have the experience, expertise, and knowledge to gear up for war. Even Allen Tough, who has written extensively on why ETs could be great benefactors (see chapter 11) acknowledges that their "gee whiz" technology is likely to include accurate and devastating weapons.[28] The bat men on the moon had no visible military technology. The engineers responsible for the canals on Mars, and whatever installed the interstellar beacons (quasars and pulsars), showed no signs of flexing their military muscles. The alien abductors of today occasionally inflict pain, but this seems to reflect nothing more than a poor bedside manner. Nevertheless, whether or not they have an overwhelming military capability, years of viewing science fiction will convince a large proportion of our population that they will be able to invade Earth and obliterate us.

"Immediacy" is a social psychological term referring to the degree of presence or involvement that characterizes social exchange. For example, communicating face-to-face, where participants are sufficiently close as to clearly see minute changes of expression, smell each other, and perhaps feel each other's body heat, would be highly immediate. Communicating through regular mail would not be at all immediate. Of course, immediacy is a continuum and not an either-or proposition. A landing on Earth would fall toward the immediate end of the continuum; a message from a civilization separated from us by immense amounts of space and time would fall toward the opposite end of the continuum. Even if they have tremendous weaponry and a hostile intent, they won't be very threatening if there is no way for them to get here. The bat men on the moon, the Martian engineers, and the interstellar navigators were locked on their own planets or safely removed from us by many light-years. The Martian invaders and the alien abductors roamed North America at will.

SETI contact scenarios involve very low degrees of immediacy. If we are able to detect anything beyond an uninformative carrier wave, it is likely to be in some sort of code that will take years to unravel if it is decipherable at all. If the message has traveled great distances, those who framed it could be long gone and the conditions they describe could have passed into history. In all likelihood, an actual meeting, if not impossible, would be generations away. There is little or no immediacy and, if for no other reason than this, little or no threat. Just to be extra safe we have started with a passive listening strategy and are somewhat furtive about our location.

The intentions we ascribe to them will make a huge difference in our perception of threat. Are they altruistic, poised to share information with Earth, perhaps eliminating poverty, suffering, even death? Are they neutral but benign, content to live and let live in the same sense as any good but somewhat "invisible" neighbors? Or will they be self-righteous, fearful, or hostile, coveting our resources, seeking to conquer us, or eliminating us as undesirables or potential troublemakers? As we saw in the preceding chapter, our assessment of their intentions will depend on many factors: what they are "really" like; the information that we can retrieve from their broadcasts; media accounts; social pressures; and our own mental processes. Suffice it to say that the accuracy of such inferences about even terrestrial societies is so poor that, for planning purposes, some military analysts focus on capabilities and ignore or discount intentions. Nonetheless, we are sure to form impressions of the kind of creatures they are and of what they want from us.

The bat men on the moon were cute, cuddly, peaceful, productive, and creative. The series of articles in the *New York Sun* detailed friendly, almost affectionate, relations not only among the bat men themselves but also among lunar species as a whole. We didn't know that much about the Martian engineers or the operators of the interstellar beacons, but their activities seemed peaceful enough. The *Mercury Theater*'s Martian invaders, on the other hand, were huge, hideous, clumsy creatures with no chins, quivering mouths, and oily "fungoid" skin. Most important, their invasion of Earth and destruction of its armies demonstrated their evil intent. The alien abductors' motives are somewhat opaque, but most of their actions suggest that they are at best rather unconcerned about human welfare.

These precedents provide at best a tenuous base for forecasting human reactions to contact with extraterrestrials. Nonetheless, they do demonstrate that there will not be one uniform reaction to all forms of contact with all kinds of extraterrestrial species. The only aliens that have generated significant apprehensions or fears are the Martian invaders and, for some people, the alien abductors. In each of these cases there was high technological development, physical proximity, and either a callous disregard for humans or open aggression.

Problem Solving

Once we have defined a situation as harmful or threatening, that is, potentially injurious to the self, we undertake a different type of appraisal. We determine whether we have the personal or environmental resources to effectively negate the threat. We try to decide if we can outfight, outwit, or outrun the predator, or otherwise control or avoid the situation. If we can defeat or escape ET, then we may feel reassured.

"Fight or flight" is the dominant reaction to threat. Our first thought after realizing that a UFOnaut is dangerous may be to determine whether we can kill it, defeat it, or at least chase it away. Several considerations may work against actually trying this: a belief in its physical or military superiority, a lingering suspicion that it might be a drunken man in a rubber suit, internalized prohibitions against physical violence, and so on. The second option, flight, involves outrunning or hiding from the creature, as many people did upon hearing the *War of the Worlds* broadcast. Interestingly, tales of close encounters often suggest a certain futility of flight. For example, it is commonly claimed that in the presence of UFOs, automobile electrical systems fail, leaving people in stopped cars, a theme inspired, perhaps, by the 1950 movie *The Day the Earth Stood Still*. Some reports of even closer encounters suggest that people who act as if they might resist or run away are paralyzed by some sort of beam or ray.

Recognition of threatening conditions coupled with a low estimate of one's power to resist results in fear. The symptoms are familiar: a rapid heartbeat, shortness of breath, weakness, trembling, and an almost overwhelming desire to be somewhere else. Reality and fear go together; the simultaneous recognition of danger and one's limited capabilities to deal with it is in the interests of survival, because it prompts protective action. Fear is normal and healthy unless it is triggered by something that is not at all dangerous or unless it is disproportionate to the real level of threat or harm.

Emotion-Centered Coping

If problem solving—fighting, hiding, talking one's way out of the situation—doesn't work, we resort to the next line of defense.

around him suddenly looked Asian and seemed to be talking in Chinese.[31]

The *War of the Worlds* broadcast prompted two surveys, a national survey undertaken on behalf of CBS by psychologist Frank Stanton (who would later become that network's president) and a study of New Jerseyites undertaken by Hadley Cantril, assisted by Hazel Gaudet and Herta Herzog.[32] These studies certainly confirmed that many people were frightened, but they also revealed some of the strategies that people use to cope with such situations. Some people attempted to verify that the story was real, and quite a few realized early on that they were listening to a play. (These reality checks didn't always work: one dial-twister found church music and, forgetting that it was Sunday, took this as confirmation that the end was near.) Because we know that the Martians had not landed we tend to be amused by people's reactions. Yet fleeing the area, trying to seal the windows against poison gas, attempting to rescue friends, and similar actions make more sense to me than simply giving up. Some people who saw no way out tried to prepare themselves psychologically through prayer or, in the case of one couple, changing into their best clothes so that they could die with dignity. Pandemonium—abandoning one's responsibilities, clogging building exits and roadways, driving recklessly, and the like—might have been minimal if it hadn't been for the sense that there was so little time. Still, some people got organized and found helpful things to do. Absent from the reports I've read of the initial broadcast are fatalities from heart attacks, suicides, hit-and-run accidents, looting, and vandalism. There were some deaths following rebroadcasts, but overall casualties were less than those following some of the worst soccer riots.

North Americans and northern Europeans in particular tend to be action oriented, and their preferred strategy is to solve problems to make them go away. Realistically, if a "War of the Worlds" occurred, there might not be all that much anyone could do. Trying to solve the problem might be less fruitful than assuaging one's feelings. Following the Three Mile Island disaster, some people tried to "fix" the problem by influencing authorities to make changes; others simply tried to make themselves feel better about living in the reactor's immediate vicinity. The authorities weren't about to be persuaded, and the people who coped internally made out better than those who tried direct action.[33] Assuming that we are not annihilated, long-term coping strategies, such as those that helped people live through concentration camps, would help us endure.

DIFFERENCES AMONG PEOPLE

Although some reactions may be more prevalent than others, under any given contact scenario groups and individuals will differ in how they react to the news. Reactions are likely to vary as a function of culture. The kinds of emotional responses that are acceptable in one culture may be unacceptable in another. Characteristics of the aliens that are pleasing to the members of one culture may be displeasing to those from another culture. We should also expect age-related differences. Discussions of human reactions to contact focus on adults and do not consider the effects on children. Children and adults will not share the same understanding of events; young children, in particular, do not think in "grown-up" ways. Finally, within a given culture or age group, we might expect some personality traits to help a person meet the news with equanimity. Other traits might make a person apprehensive and negativistic. All of this means that even the most clear and detailed announcement will elicit a variety of responses.

Sensation or Thrill Seeking

People differ in their appreciation of new and unusual experiences. "Sensation seeking" refers to people who get particular pleasure out of risk, novelty, and excitement.[34] Compared with people who are somewhat more conventional, sensation seekers like to jump out of airplanes, climb cliffs, experiment with mind-altering drugs, develop unconventional relationships, and in other ways explore the new and different. We might expect that compared with more conventional people, who are somewhat averse to risk-taking and prefer the familiar and comfortable, sensation seekers will respond more positively to the discovery of intelligent extraterrestrials.

Although it might seem that sensation seekers would be good choices for deep-sea diving, exploring Antarctica, voyaging to outer space, or joining a group of envoys to negotiate with emissaries from a different solar system, they would not. Hollywood portrayals to the contrary, such situations are risky enough without people who enjoy taking chances making them even more dangerous. Consequently, sensation seekers might delight in the news of contact with an extraterrestrial society, but let's hope they are not the ones who will actually "meet" ET.

Tolerance for Ambiguity

Initial reports are likely to be quite sketchy, and years could pass before the two cultures find ways to trade meaningful information. Thus, people's initial responses are likely to be based on incomplete information. People differ in their ability to tolerate ambiguous situations, that is, to live with uncertainty or a lack of closure.[35] We can expect people who have a high tolerance for ambiguity to respond more positively than people who are uncomfortable when there are gaps in their knowledge or when they must contend with conflicting pieces of information.

Dogmatism

People differ in terms of the flexibility of their opinions. Dogmatic individuals have high confidence in their views of nature and the universe, and a clear, unshakable view of right and wrong.[36] Furthermore, they believe that their own views should be held by everybody else. They tend to see people who don't agree with them as stupid, ill-informed, misled, or perhaps even duplicitous.

Dogmatic views are rigid views, and psychologically, people

with dogmatic views try to maintain them at all costs. Such views may include a very orderly and conventional view of the universe, absolute standards of right and wrong, and strict adherence to fundamentalist religions. Contact with an extraterrestrial society will force many people to do a new take on the nature of the universe. This would be particularly challenging for dogmatic people.

Some people who believe that humans are unique, as they have been created by God in God's image, may have difficulty reconciling this belief with the discovery that there are highly advanced beings elsewhere. For some such people, humankind occupies a unique place in the universe and has a special relationship with God. The difficulties may be particularly pronounced for adherents of religions that posit that Earth has been visited by one or more special emissaries who have not visited other corners of the universe. For members of sects founded upon a particular sequence of events (such as Christ's visit to Earth), the discovery of worlds where such visits did not take place could strike a dissonant chord (see chapter 11). Rather than assimilate the discoveries, they may respond with greater loyalty to their sects and a stronger adherence to their preexisting views.[37]

Repression and Sensitization

You are walking down the street at night and you hear loud footsteps rapidly overtaking you. Repression and sensitization are two techniques that you might use to cope with this situation.[38] "Repressors" tend to ignore the possible danger, that is, to keep it out of conscious awareness. By ignoring the footsteps, telling themselves that it is probably just a police officer walking a beat, and so forth, they maintain psychological comfort. The other strategy is "sensitization," that is, heightened alertness to the situation, which prompts an energetic search for protective action.

Repressors, I expect, would greet the news of contact with relative calm and equanimity. Sensitizers would be very aware of the potential dangers. In their extreme forms, repression might result in a failure to react to dangerous conditions; sensitization, in an overreaction to harmless conditions. In other words, extreme repressors are the ostriches of this world, and extreme sensitizers are the Chicken Littles.

Emotional Stability

It has long been expected that people who lack emotional stability will be "at risk."[39] The idea is that people who are able to hold jobs and maintain decent relationships but who have few reserves to cope with special stresses may break down on hearing news of contact.

There is at least some basis for this concern. The effects of stress tend to be cumulative. People who are already under a great deal of stress due to health, family, or professional difficulties are unlikely to react well to additional stress from the discovery of extraterrestrials. Furthermore, the ability to cope with stress is, in part, a stable property of individuals;[40] on this basis we would expect that people who have shown an ability to adjust to stress prior to contact will do better than people who have a record of unsuccessful coping.

We might also expect ideas about our newfound acquaintances to enter into the psychological dynamics of people who already have mental health problems. Delusions and hallucinations reflect both the time and the culture, and it would be surprising indeed if the discoveries about extraterrestrials did not appear in delusional systems.

Affectivity

Some people tend to be positive, upbeat, and cheerful, whereas others tend to be negativistic, anxious, and depressed. Although all but the most cheerful person can occasionally be discouraged (and all but the biggest grump is uplifted on occasion), people tend to have fairly stable levels of optimism and cheerfulness.[41] A person characterized by positive affectivity tends to look on the bright side of things—every cloud has a silver lining. A person characterized by negative affectivity tends to dwell on the downside of things—every silver lining hides a cloud. Since affectivity influences so much else about a person's life, we might expect it to color the reactions to extraterrestrials.

Locus of Control

"Locus of control"—internal or external—relates to a sense of being in charge (or not in charge) of one's own life.[42] People who tend toward the internal end of the scale see themselves as able to affect conditions in such a way as to achieve outcomes that are favorable to themselves. "Internals" view themselves as acting on the environment, rather than being acted on by the environment. People who tend toward an external locus of control view themselves as relatively powerless; the outcomes they experience are attributed to forces external to the self. "Externals" view themselves as acted on, rather than as acting. In a sense, locus of control separates the "can do" people from the defeatists. Locus of control sits on a continuum; very few people see themselves as totally in charge of every situation or as hapless victims of circumstance.

"Internals" tend to do well in many areas of life because their faith in their ability to get results prompts them to try. I would expect that people who tend toward an internal locus of control will react more favorably to news of contact than people who have an external locus of control, because the "externals" are likely to see the aliens as one more unpredictable force running their lives.

Self-Esteem

Self-esteem is a judgment about one's own worth.[43] High self-esteem means feeling good about oneself, one's goals, and one's capabilities. People who have high self-esteem are better positioned to take threatening circumstances and minor setbacks in stride. Whereas on occasion we all fail or are embarrassed, people who have high self-esteem possess a fundamental sense of worth coupled with confidence in their own abilities and their future. We might expect that in comparison with people who have low self-esteem, people who have faith in themselves and a sense of being "OK" will be better able to accept news of superior beings elsewhere. For those with low self-esteem, advanced beings will offer yet another benchmark that can't be met.

CONCLUSION

SETI searchers hope to provide the public with prompt, accurate, and detailed information on the detection of alien life-forms, once the finding is confirmed. According to their plans, at least, we may expect a massive information campaign and strenuous efforts to minimize confusion and rumor. Their expectation that contact will be a positive moment in human history, coupled with their efforts to educate the public, should encourage a positive reaction.

Nonetheless, we cannot, with complete confidence, predict people's initial reactions to the discovery of an extraterrestrial society. Our reactions will depend on who the aliens are, what we know about them, their nearness or immediacy, and our perceptions of their capabilities and intentions. Moreover, under any given contact scenario, reactions are likely to vary as a function of culture, age, and personality.

There are five historical precedents when large numbers of people believed that intelligent extraterrestrials have been discovered. These include the discovery of bat men on the moon, the observance of canals on Mars, the *War of the Worlds* broadcasts, the identification of quasars and pulsars as being of intelligent extraterrestrial origin, and the UFO sightings and alien abductions of today. (Of course, not everyone bought into these false alarms.) From what I can tell, the predominant reactions to the bat men on the moon and the canals of Mars were anticipation and pleasure. The predominant reactions of those who even heard of quasars and pulsars may have been subdued excitement. Thus far, only the *War of the Worlds* broadcast, which depicted a live confrontation with a hostile and superior military force, resulted in panic.

Everything in SETI is a matter of probabilities. It is highly unlikely that contact will be made through the arrival of alien invaders on our planet. Contact, if it occurs, is likely to occur through the interception of microwave radio waves, and the information they carry is likely to be difficult to decipher or not particularly exciting. There are many reasons to believe that, apart from the initial flurry of curiosity and excitement, this contact will be, for most people, not particularly frightening. As David Swift has pointed out, because contact is likely to result from the deliberate efforts of scientists here on Earth, it will not be a surprise; contact is taken for granted by a large part of the population; and most people have more pressing things to worry about.

To some extent, many people are already prepared. Surveys conducted over the years show an increase in the percentage of respondents who believe that intelligent extraterrestrials exist, and the more recent surveys place the percentage of believers at about 50 percent. Although relatively few studies have explored people's attitudes toward communicating with ET, the limited available data suggest that most respondents view communication as beneficial (or at least, nonthreatening) to humankind.[44] Thus, suggests Swift, "the safest single-sentence prediction about the probable effect of contact on daily life is that it will be minimal."[45]

TEN

BUILDING RELATIONSHIPS

Shirley Varughese offers sensible advice to anyone who happens to encounter an alien. First, if it moves in your direction, maintain eye contact and walk backward, but slowly, so as not to startle the intruder. Second, if you have a weapon or a weaponlike thing, such as a flash camera, put it down. (A flash of light could be misinterpreted as a death ray or laser-type weapon.) This would be seen as a friendly gesture. However, if you don't have the confidence to lay it aside, place it across your chest "or let it hang down to your side, ready but not aimed. This would be a clear message to anyone who did not speak our language."[1]

These precautions were formulated when the United States was actively involved in the manned exploration of space. The scenario revolves around astronauts accidentally encountering another intelligent species while exploring our solar system. This could happen without warning, and the recommended precautions (reminiscent of those that one might take when approached by a wild animal or an unfamiliar dog) were intended to avoid getting off to a very bad start.

Under a SETI-type scenario, contact is a bit less dramatic, as it occurs through the interception of radio waves. As foreign service officer Michael Michaud points out, once we have identified a radio transmission as being of extraterrestrial origin, we may think we have a choice.[2] We can remain silent, hoping that they won't notice us, or we can try to initiate a dialogue. The former option is not really an option at all; if our radio transmissions speed outward, our presence becomes known. The second option—learning about one another, seeking a common ground, and establishing mutual priorities—could be the only realistic one.

WHO'S IN CHARGE HERE?

Diplomatic contact of any type involves not only individuals who are in key leadership positions but also agencies and institutions. Indeed, Varughese warns astronauts not to try to engage in negotiations with the aliens but to go through the proper channels. Given this, it makes sense to consider how certain aspects of human organizations could affect the course of interstellar affairs.

Jurisdiction

A radio interception could occur anytime, anywhere, but it is far more likely to occur in a wealthy and scientifically advanced country that has both the equipment and technical skill to intercept weak signals and determine what they are. Under a SETI-like scenario, the first human to encounter alien intelligence is likely to be a well-to-do, highly educated scientist who has already given the matter some thought. A spacecraft, on the other hand, could land in any nation on Earth, in a neutral territory such as Antarctica, or in the ocean. The first contactee could be almost anyone, including someone who has *not* given the matter much thought.

Just for fun, let's suppose an alien spacecraft landed in New Mexico and its occupants demanded to be taken to our leader. Who might that be? The sheriff or mayor? The governor of New

Mexico or the president of the United States? The secretary-general of the U.N.? In the off-chance of an in-person landing, local officials could leave an indelible stamp on interstellar relations, simply because they are likely to be the first on the scene. Funny science fiction stories take an unkind view of local authorities as hopeless bureaucrats intent on mindlessly enforcing local laws. The list of laws and regulations that might be broken by extraterrestrial interlopers is a long one, including use of unregistered vehicles, trespassing, and creating a public nuisance. Scary science fiction stories present local authorities as stupid, foolish, trigger-happy individuals who almost immediately initiate a fight. This reflects the troops' conviction that the aliens are dangerous, coupled with a supreme sense of confidence in their weapons. As science fiction readers know, the misguided warriors who do this invariably underestimate the "invaders," and annihilation is the usual reward.

Research on aggression educates our guesses about the conditions under which a human team might provoke a military confrontation.[3] One contributing factor, not too surprisingly, is discomfort or pain. The physically damaged individual tends to lash out even if it is not clear that the intended victim of the aggression is responsible for the pain.[4] A second factor is the presence of aggressive cues—angry facial expressions or gestures, military uniforms, weapons of any sort, signs of strength, or violence.[5] And, of course, aggression by one party is likely to touch off aggression by other parties. There is some truth to the idea that if someone opens fire, everyone else will start shooting, too. But, most important, the authorities on the spot are likely to be under extreme stress and intense time pressures, conditions that do not encourage a level-headed and even-handed response.

Many state, national, and international agencies will have a stake in the situation. Obviously, the military will be concerned about any potential threat. Certainly, the air force will be sensitive to the accusation that alien craft have invaded the skies. The navy will be interested if they fly over water (which is likely given the distribution of land and water on our planet), and the army might jump in if they set foot on land. Judging by past government interest in UFO phenomena, we can expect involvement on the part of various security agencies, including the National Security Agency, the CIA, and, of course, the FBI.[6]

The Department of State will be involved to the extent that we view alien societies as nation-states. Certainly, Customs and the Immigration and Naturalization Service would take note, as would any other agencies concerned with the passage of personnel and materials across national boundaries. The Federal Aviation Administration will fret about use of national airspace, the enforcement of safety regulations, and, perhaps, aircraft landing in undesignated locations. (On the other hand, maybe those who fly such craft will be safety conscious: an emerging theme in UFO reports is that some saucerlike flying objects now display the familiar pattern of flashing lights that conventional aircraft use for safety purposes. This discovery is interpreted either as evidence that the saucerlike vehicles are under human control or as clever attempts at camouflage on the part of nonhuman beings.)

Jurisdiction could depend, in part, on perceptions of the ETs. Are they humans, with human rights (whatever that might mean) or are they animals? Are they visitors or immigrants, in which case Customs and Immigration would come to the fore, or are they invaders, which would require a police or military response?

Given the less exciting but more probable SETI-type scenario, there should be time to alert the proper authorities and prepare a reasoned response. SETI scientists and enthusiasts argue strongly against local or national responses, as these are likely to be premature, ill-conceived, or inconsistent with the welfare of all humankind. The detection of extraterrestrial intelligence, a momentous event for everyone, demands a response from the world as a whole.

Ernst Fasan finds a firm basis for supranational action in international law.[7] Central here is the 1967 Treaty on Principles Governing the Activities of States in the Exploration and Use of Outer Space, Including the Moon and Other Celestial Bodies. According to Fasan's interpretation of this binding international agreement, space exploration is for the benefit of all humankind, astronauts are "envoys" of Earth, and each nation must inform all other nations of activities and discoveries that might constitute a danger. The 1973 Convention on International Liability for Damage Caused by Space Activities specifies that if one nation's space activities damage another nation, the latter nation may hold the former nation liable. Consider the damages that might result from withholding evidence of extraterrestrial intelligence. Finally, the 1979 Agreement Governing Activities on the Moon and Other

Celestial Bodies recognizes that extraterrestrial life might be discovered, that Earth is worth preserving, and that extraterrestrial life might pose dangers to humankind. All these documents develop the themes that space exploration should be done responsiably and that exploring nations must consult with other nations that might be affected too.

As Stephen Doyle points out, a world forum will be useful: it could conduct an accurate and thorough investigation of the contact and establish the scientific, military, and political realities; ensure that people are fully and accurately informed; debunk false or exaggerated claims; bring special-interest groups into line and discourage unilateral action; and, if warranted, guide a confident, orderly, integrated world reply. Although this would require many different types of institutional arrangements, the United Nations should play the leading role in the stabilization process.[8]

There are practical as well as legal justifications for coordinated international action. Observatories from around the world will be needed to verify the signal. Furthermore, it will be useful to draw on experts from many different countries to decode the signal and help prepare a suitable response. A supranational response is generally a higher-quality response (neither too submissive nor too aggressive, for example) and would in any event prevent the discouragement that an extraterrestrial society might experience if it were bombarded by an incoherent package of conflicting messages beamed by many individual governments. Joint actions should calm fears that the world political system will be thrown into instability if one nation gains an advantage by communicating unilaterally or by forming a special, perhaps unique, relationship with an advanced civilization. The preparation of the response, then, requires careful planning and international unity.

The February 1990 issue of *Acta Astronautica* offers a Declaration of Principles concerning Activities Following Detection of Extraterrestrial Intelligence.[9] This declaration is intended to be binding on the signatory individuals and institutions participating in SETI. The protocol begins with an agreement to verify that extraterrestrial intelligence is the most plausible explanation of the candidate signal or other evidence and then to promptly inform observers at other sites so that the signal's interception can be replicated. Following verification and replication, the secretary-general of the United Nations and a long list of scientific institutions and organizations should be notified. Scientific communications channels and the mass media will disseminate the news widely. The radio frequencies that carry the transmission would be protected, and there are ongoing procedures for data collection, verification, and analysis. There will be no reply in the absence of "appropriate international consultation." All of this will be subject to oversight and refinement by the SETI committee of the International Academy of Astronautics, in cooperation with the International Astronomical Union, an organization that has the means to rapidly disseminate information to observatories throughout the world.

Allan E. Goodman, of the School of Foreign Service at Georgetown University, has discussed the responses of some scientists and policy makers to an earlier but similar postdetection protocol. Based on a 1971 declaration by U.S. and USSR scientists, this draft protocol included as key elements the free and open dissemination of information, and international consultation for analyzing the signal and formulating a response. Respondents offered several arguments against such a protocol. Some questioned the necessity or practicality of preparing it: extraterrestrial intelligence might never be discovered or the scenario could differ so much from our expectations that the efforts would be in vain. Furthermore, critics suggested that under a SETI-type scenario there should be plenty of time to develop protocols after contact occurred.

Some of Goodman's respondents expressed concern that such protocols could infringe on national sovereignty or academic freedom. Shouldn't individual nations be allowed to benefit from detecting extraterrestrial intelligence on their equipment or through their recovery of alien artifacts? Shouldn't individual scientists be free to choose the best way to follow up on their startling discoveries? Another concern was that dissemination of the news, especially rapid dissemination of the news, could lead to false alarms and professional embarrassment, trigger panic, or undermine national security.

Goodman takes these objections seriously but does not consider them insurmountable to the implementation of an international SETI protocol. He argues that, even in times of international tension, it will be in national interest to share information about

SETI, even as one might share information about volcanic eruptions, earthquakes, and disease. The fact remains, however, that some messages have been sent into space without international consultation "and even by some entrepreneurs who offer to beam up 25 words of a message to the stars into space for $9.98."[10]

Donald Tarter proposes there be an immediate response to any contact. He notes that most SETI scientists will want to reply. "A discovery with no response," he observes, "would seem to most scientists to be leaving the most exciting discovery in history in an incomplete stage of investigation."[11] If we wait to formulate a supranational reply, the decision will be made under intense media coverage. In the meantime, some people may attempt to arouse public fears and others may lobby for a return transmission that serves their personal interests. The SETI protocols are agreements, not laws, and many countries or even organizations could send unilateral responses. The first recognizable message that ET receives may not serve the interests of all humankind, but rather the special interests of one or more fringe elements. Furthermore, even as the first information that we have about them is likely to have a disproportionate impact on our perceptions of them, the first information that they have about us is likely to play a weighty role in their perceptions of us. It could be very difficult for a later transmission, sent on behalf the entire world, to correct any misperceptions.

Tarter hypothesizes that, apart from identifying the first signal from us as the "official response," they will tend to place more trust in messages that come from the most powerful transmitters. He urges us to reply immediately, using the best available transmitter. Our first signal should acknowledge our receipt of their signal, establish itself as our planet's official response, and contain a secret code that could be used to identify future official transmissions. ET would be encouraged to ignore or discount transmissions that did not contain the code, as they would carry unauthorized or renegade messages.

The protocols I have discussed thus far are based on the detection of a signal from light years away. As Allen Tough points out, they do not apply to other situations, such as the detection of a small robot probe. Whereas some of the principles set forth in preexisting protocols would apply to probes, others would not. For example, whether we detect a remote signal or a probe, it makes sense to confirm or disconfirm its extraterrestrial origin and then make the data widely available. On the other hand, the specific authentication procedures set forth for validating a radio signal would not easily translate to a probe, and the lengthy process specified for formulating a response to a radio interception might need to be curtailed if the probe requires an immediate response. Tough recommends the formation of an "authenticity team," including experts and intellectual leaders from a wide range of disciplines. This team would be empowered with sufficient freedom and flexibility to examine the probe and, if necessary or appropriate, to respond immediately.[12]

Bureaucratic Responses

Most government agencies, including those that would deal with extraterrestrials, are organized along bureaucratic lines. Despite near-universal condemnation, the bureaucracy is, under many conditions, a rational and efficient form of organization. Unfortunately, bureaucracies tend to perform poorly under the conditions that we would expect to accompany many contact scenarios.

Bureaucratic decisions tend to be based on precedent, and operations follow standard procedures. Confronted with a situation that requires a response outside the organization's established range, bureaucratic organizations run into trouble. The old ways are applied to the new situation, even though the old ways are not really applicable. Certainly, dealing with intelligent extraterrestrial life-forms will be well outside the experience of our bureaucracies and will require something other than a preprogrammed, stereotypical response.

Then there is the slow pace at which information moves through a bureaucracy. For example, if a UFO managed to collide with a navy airplane, the report would work itself up the line through the chief of naval operations to the secretary of the navy and to the president. The official response would then wend its way downwards. Meanwhile, the news media would have long since issued the first public account. (To help offset this problem, some Pentagon employees, both military and civilian, currently monitor the national news networks so that public relations officers can be briefed before they hear official word of an incident.)

Yet another problem is that people in key positions lack crucial information because it is filtered out before it reaches them. Subordinates are reluctant to present information their superiors won't like; for example, information that is unflattering to the boss or agency suggests that the agency is facing big problems, or hints of an impending failure. The censorship of important but distressing information is the greatest in large organizations with many hierarchical levels.

Still another problem, according to Kenneth Boulding, is that people who are high in organizations frequently have an exaggerated sense of their power, especially relative to a real or imagined enemy.[13] There are at least two reasons: information that runs counter to self-aggrandizement is censored as it moves up in the hierarchy; and it is the power-oriented individual who has a better chance of attaining high-status positions. An overestimation of power relative to ETs could encourage a predisposition to respond with threat and force.

Further difficulties arise when we consider that, in group decision-making situations, critical thinking is undermined by pressures to preserve social harmony within the group. Irving Janis coined the term "groupthink" to designate this process.[14] In groupthink, discussions focus on a few alternatives, in particular, those that are initially favored by "the boss." People who have contradictory thoughts tend to keep silent so as not to upset other members of the group (even though many of them may harbor the same thought). Dissenters who do speak out are contradicted by "mind guards." The result is that the group focuses on a narrow range of alternatives and makes decisions based on incomplete, faulty, or erroneous data. Characteristically there is little discussion of ill omens or of the kinds of things that might go wrong. Perhaps consistent with this is John Keegan's observation that military intelligence analysts tend to form a hypothesis early on and stick with it even in the face of growing counter-evidence.[15]

Under certain contact scenarios (such as those involving ominous aliens or requiring a near-instant response) people who have to make decisions may be under great stress, and stress itself can work against sound decisions.[16] First, people under stress tend to focus their attention on a single option without carefully analyzing its pluses and minuses or evaluating it in light of potential alternatives. Stress makes it easy, in other words, to latch on to the first or second plausible plan that comes along. Second, stress sometimes leads to "defensive avoidance," that is, finding ways to avoid the problem. Defensive avoidance is likely when one has come up with some alternatives but none of them seems quite right. It includes wishful thinking, denial, rationalization, procrastination, buck passing, and other ways of evading the issue. Finally, when people believe that they have little or no time to find a reasonable solution, they are likely to engage in panicky, ineffective decision-making. Because all these conditions undermine the quality of the decision-making process they typically lead to poor decisions.

NEGOTIATION STRATEGIES

Power is the means to preserve or exercise options, to get one's way. According to Boulding, whether we are talking about individuals, organizations, or societies, we use basically three strategies for getting what we want: threat, economic power, and integrative power—roughly, the stick, the carrot, and the hug.[17]

Threat involves scaring or coercing others with the promised or actual application of destructive force. This is the foremost power of the police, the military, and many regulatory agencies. Economic power involves bargaining or trading; it is the basic process of the marketplace. Integrative power, roughly synonymous with love, is based on legitimacy, respect, friendship, and other centripetal forces that underlie many fraternal organizations, political parties, and religious groups. Integrative power is inherent in our recognition that other people or institutions have claims upon us.

Although in any given situation one type of power may predominate, the three forms coexist and interplay in complicated ways. For example, even armies whose purpose is to intimidate potential opponents also have an integrative component. This is reflected in the practices and rituals that reinforce a sense of identification with the group. Troops are encouraged to take pride in their unit and to look out for one another and maintain a sense of family (exemplified, perhaps, in the U.S. Marine Corps' motto, Semper Fi). Although religious organizations are based on integrative power, sometimes they use threats: if you break our rules you will go to hell (or at least be kicked out of our group).

What strategies are available for us humans to use in our negotiations with extraterrestrials? What types of strategies are available to them? Are some strategies more likely to be used or more likely to be successful than others? Continuing to follow Boulding's analysis, we can group strategies according to their emphasis on the stick, the carrot, or the hug.

Threat-Based Strategies

Destructive force serves constructive purposes when, for example, we use steam shovels, bulldozers, and dynamite to clear land for construction or to rupture the Earth's surface to extract material to form useful artifacts. But we are most familiar with destructive power used in destructive ways. Individuals get into fights with one another, organizations attempt to put one another out of business, and nations posture and go to war.

Societies place limits on the exercise of destructive power, but these limits vary from society to society and are not always effectively enforced. The belligerent or violent person is physically subdued, medicated, or thrown into jail. The destructive organization is subject to legal regulations and industry or governmental sanctions. Supranational systems such as the United Nations try to impose constraints on individual nation-states. The use of force is constrained by higher-order systems: the individual is constrained by the organization, the organization by the society, and the society by international law.

Given our presumed technological inferiority, threat and force do not loom as promising tools for exacting concessions from our new acquaintances. In human history, the most lopsided battles took place in the 19th century.[18] On the one side were the British and other northern Europeans equipped with steam-powered ironclad gunboats, breech-loading repeating rifles, and, later on, machine guns. On the other side were Africans and Asians armed with knives, spears, primitive cannons, and ornate but inferior cast-iron flintlocks that tests showed would misfire almost as often as not. Thus, chunky British gunboats pursued elegant Burmese war boats until the latter's oarsmen were too tired to continue. A French force of 320 destroyed a Sudanese force of 12,500, and a group of 20 Britons accompanied by 20 Egyptian allies eliminated 11,000 dervishes before their own lives were taken. Of course, European technology was only a century or two ahead of their opponents', whereas we expect the extraterrestrials' technology to run ahead of ours by millennia. Could they knock our planet out of orbit, drop a comet on the White House, or turn our sun into a supernova?

It is worth noting that even with superior technology and power, large terrestrial countries sometimes get in trouble when they send troops to places they don't understand to fight people whose cultures they don't know—places like Afghanistan, Bosnia, and Vietnam. The challenges posed by an imperfect understanding of human enemies should pale into insignificance relative to those that would arise if we were to pit ourselves against alien life-forms, especially outside our own turf.

Many problems might ensue if human troops were assigned to combat in outer space.[19] Most obvious, of course, are logistics and supply difficulties. Weapons systems developed on Earth would have to be redesigned or at least recalibrated to operate in worlds with different gravities and atmospheres. (For example, artillery range would be extended dramatically on the moon, where the shell would have to overcome but a fraction of the Earth's gravity and there is no atmosphere to slow it down.) Certain terrestrial weapons might not be at all effective against certain kinds of organisms; for example, a self-sealing gas bag would not be deterred by bayonet jabs or bullet holes. Then there could be psychological barriers to overcome: withholding fire on scary-looking but kind creatures, attacking cute and cuddly but dangerous creatures. The basic principles of war might have to be rewritten to deal with new enemies in new situations. We might want, for example, to lure creatures from oxygen-rich planets into the mountains; to lure creatures from oxygen-poor planets into the suffocating atmosphere at sea level; to attack mole people during the day, and those that need a lot of light, at night. Advanced technology may not be the only reason that death rays, thermonuclear weapons, and the like typify discussions of the military in space. Space warriors may need such weapons because they are capable of broad-spectrum destruction and should be effective against a variety of life-forms on many worlds.

Our fascination with superweapons should not obscure the fact that the nature of warfare varies from culture to culture. Organized armed conflict is not invariably extreme and ruthless; many cultures impose constraints on war-making. Limitations include the exemption of certain persons from combat (the young, the old, the infirm); conventions governing the when, where, and how of combat; and ritual, which insists that combat must take a certain form and dictates at what point arbitration and reconciliation must occur. Unlike our own societies, many so-called primitive societies had many means to limit the havoc they wreaked upon each other, and John Keegan suggests that restrained combat may also be the way of the future.[20]

Given the likelihood of their huge military advantage, can we expect extraterrestrials to use force to subjugate us? Boulding points out that threat is costly to the oppressor as well as to the oppressed. First, there is the expense of developing and maintaining the enforcement mechanisms: the military infrastructure, intelligence systems, weapon systems, and so forth. Then there is the cost of maintaining and deploying the troops. All of this diverts money from activities that enhance the welfare of the society. Boulding also argues that the nations that seem to have done best in the last hundred years or so are those that have stayed out of war. Perhaps, as suggested in chapter 7, other members of the Galactic Club will have had the same experience.

Losers have a knack of recovering, economically and politically. Nations that are ripped apart tend to reunite. Or the invaders are ingested within the land of the vanquished. For instance, foreigners who invaded China over the centuries have either been booted out or themselves become Chinese. Given the costs of war and the resilience of the "losers," perhaps most neighboring civilizations will have discovered that there are better ways than force to get results.

Economic Strategies

Economic power is based on the concept of exchange. This is a buyer-and-seller type of arrangement. To use this type of power, one side offers something (matter, energy, information) that the other side wants. As long as each side sees itself as treated fairly, the relationship is likely to flourish.

The question is, in a bargaining relationship, what might we offer them? Clearly, they would have much that could be of value to us, but would they have everything? Would trying to find something to please them be like trying to find a present for that rich relative who has everything?

Here we would do well to remember Frank White's point that development will proceed at differential rates across different domains.[21] There may be some areas where we have a relative advantage, and we should avoid too narrow a view of what we have to offer. For example, rather than become obsessed with our mediocre technology, we might consider our strengths in literature and arts. Terrestrial societies that vary widely in terms of their technological sophistication can still appreciate (and patronize) one another's arts.

Among human societies, a situation in which each side gives something and receives something is much more comfortable than a situation in which one side gives and the other side receives. The side that always gives tends to feel used, and the side that always receives feels guilty. Both sides tend to become hostile. Reciprocity need not always be immediate, nor must it invariably be of equal magnitude, but nonreciprocal relationships usually fail. This is evident in everything from personal favors to international relations.

Pertinent here is a vast body of research based on the Prisoner's Dilemma Game.[22] In this game, players can either cooperate with each other and each achieve modest gains (win–win), or compete in a quest for high personal gains at the other player's expense (win–lose). However, if both players act competitively in an attempt to exploit each other, they both lose (lose–lose). Despite the initial attractiveness of the competitive response, an exploited partner will soon retaliate, locking the two players into a string of mutual losses.

There are hundreds, perhaps thousands of studies involving this game and its variants, as well as the naturalistic situations that preserve the game's formal properties. The evolution of win–win strategies has been shown in games involving animals, people, organizations, and societies. Cooperative strategies gain toeholds even in competitive environments, rapidly spread, and, once entrenched, protect themselves from invasion by less cooperative strategies. Cooperation among animals shows that the win–win pattern does not require foresight, but in the case of "higher animals" such as humans, the likelihood of cooperation is greater if the players believe they will meet again. From a strategic point of view, the most successful strategy is to start out "nice," that is, unconditionally cooperative, but to retaliate if the other player defects.

In the aggregate, analyses of situations that have the formal properties of the Prisoner's Dilemma Game cut across species, cultures, and historical epochs. Could it be that the evolution of cooperation is truly universal? If so, when one extraterrestrial society encounters another, it may be with the hand of friendship extended, but with a firm determination to retaliate if the other civilization fails to reciprocate. Perhaps extraterrestrial leaders would do well to consider the rules for succeeding at the game: "Don't be too envious, don't be the first to defect, reciprocate both cooperation and competition, and don't be too clever."[23]

Integrative Strategies

Integrative strategies rest on the forces that bind together people, organizations, and nations. They revolve around shared perceptions and common beliefs, values, and symbols. The powerful person, organization, or society is one that can communicate, persuade, create loyalties, and build social bonds. Power is being able to create and communicate a vision of the future, and then get others to "buy into it," whether this vision involves a new generation of computers, a new path to eternal salvation, or a better social order.

The world's great religions are based on shared visions, and so is the scientific community. The world scientific community transcends international borders and welcomes all, regardless of race, creed, or religious belief. The shared vision is the love of truth. Appeals to authority and force are anathema to scientists. Scientists who attempt to resolve issues with force rather than reason or who are caught deviating from the truth (i.e., fudging data) lose the respect and friendship of their peers and are effectively forced out of the community. Politics does play a role, for example in the funding of scientific research, but in principle, funding is based on agreed-upon criteria of merit. Perhaps another shared vision would be a stable, peaceful international political system, such as described in chapter 7.

Boulding argues that, despite the spectacular nature of destructive power, it is integrative power that gets results. From his analysis we might deduce that our new acquaintances will be inclined toward integrative strategies. He notes, for example, a trend away from warfare, the growth of systems to constrain violence, and a trend toward supranational entities. He argues that historians tend to assign far too much weight to warfare and that warfare is *not* a mainstream activity. It has consumed, at most, 10 percent of human time and energy.[24] Michael Hart has suggested that whereas on Earth there is a major war every 50 years or so, in space we might expect one every 50,000 years.[25]

Social systems based on integrative power last longer than those based on threat and coercion. It is not the world's greatest armies that have survived the longest, but the world's greatest religions. This is consistent, of course, with the idea that unstable and warlike societies will blow themselves up, whereas stable and peaceful societies will continue long enough to mount a sustained search for civilizations such as ours.

Finally, consider a continuum with the threat of destruction at one end, economic exchange in the middle, and integrative activities at the other end. Boulding argues that as we move from the malevolent to the benevolent end of the continuum the exercise of power rests less and less on matter and energy and more and more on information. Under a SETI scenario, interstellar relationships will be based on the exchange of information. The more malevolent forms of power, which are based on transformations of matter and energy, will be useless, while the more benevolent forms of power, which are based on transformations of information, will remain in force. Call me names or hurl threats, and I just might turn my radio off. Present me with a vision that I like, and I just might stay tuned.

GETTING STARTED

Foreign service officer Michael Michaud argues that while we hope we will contact a benevolent society, we must be prepared for something else: an empire bent on expansion, a species that has encountered nothing but hostility and aggression in its explorations of the universe, or a dogmatic species whose belief systems or societal values are threatened by the discovery of humankind.[26] Both societies, ours and theirs, will recognize this. Given the potential risks, Michaud sets three priorities for humankind.

First, we must guarantee our safety and security. This might involve agreeing not to enter each other's solar system, or establishing some form of arms control. Of course, negotiations will be difficult due to communication problems: time delays (perhaps exacerbated by a species that lives forever and has no concept of "rush"), the lack of a common language, few or no common reference points, and cross-cultural misunderstandings.

Second, we need to develop, or gain entry into, a stable interplanetary political system. To these ends we should establish diplomatic relations, whether this would involve exchanging ambassadors or establishing secure, official radio communications links. Interaction should be limited until the proper protocols are put in place. Diplomatic protocols serve to defuse situations and to minimize the chances of conflict, particularly conflict based on personality clashes, emotions, cultural differences, and rapid, unthinking responses. Diplomatic relations would make it possible to move into all of the areas associated with supranational regulation: resource consumption, transportation and communication, waste management, safety and health, and trade (see chapter 7).

Third, we should seek scientific, technical, and other information that will help us advance our own society. This third priority, tapping the extraterrestrials' immense wisdom, is oftentimes placed first, but it is meaningless if our own security is not guaranteed.

Michaud suggests that our best initial stance is to treat all communicating nations as equals and for each nation to maintain a low profile. Extreme caution is called for, given both our ignorance of them and the assumption that they will possess superior technology.

Barbara D. Moskowitz offers two guidelines.[27] One of these is the doctrine of free choice, that is, a "live and let live" policy, by which we do not attempt to impose our value system on them. The other is reciprocity, also known as the Golden Rule. This states, simply, "Respect others in the manner that you yourself would like to be respected, and act accordingly."

Anthropologists such as Ben Finney remind us to try, as much as possible, to avoid an ethnocentric stance.[28] The customs of a particular society should be accepted as they stand and interpreted within the framework of that society's culture. There are no absolute rights or wrongs; everything is relative to the culture within which the value exists or the activity occurs. An evaluative or ethnocentric stance sets the stage for conflict by communicating perceived superiority of one's own group and inferiority on the part of the other groups, leading to invidious comparisons that have the potential of provoking defensiveness or aggression.

Several of these themes are evident in Ernst Fasan's independent analysis, which leads to recommendations for the actual content of Earth's initial response.[29] Fasan believes that certain qualities are likely to apply to both terrestrial and extraterrestrial intelligent life-forms. According to him, aliens, like humans, will seek to preserve their own life and the life of their species, protect themselves from intrusion and damage, and possibly expand their own realms. Thus, we indicate that we will not harm them and we will not allow them to harm us. If one society harms the other, then the offending society offers restitution. We regard them as our equals, and we expect that they will regard us as equals. Fasan's list continues, but the central themes are equality, a live-and-let-live policy, and an emphasis on reciprocity.

There are many interesting questions regarding the psychology and social dynamics of any human team that might enter into negotiations. One of these, of course, is how we would identify people who are likely to be effective negotiators. Certainly high intelligence and emotional self-control will be important. Whereas we might wish for people who can maintain a tough stance, we will want to disqualify sensation seekers and people who like to take chances. Of overriding importance will be open-mindedness and an ability to get beyond earthnocentric, ethnocentric, and egocentric modes of thought. The composition of the overall team is also important; here we might note that in addition to scientists and diplomats it should be useful to include biologists, psychologists, anthropologists, sociologists, economists, and political scientists who can help the team understand the biomedical, behavioral, and social implications of the negotiations.

We should assume that the team will operate under relentless stress. Certainly, the hours will be long and the work difficult. Even if progress is glacially slow, the stakes could be very high and the negotiation team could be under constant media attention. But more important, the human team (and its alien counterpart) will serve a delicate boundary-spanning function. For each team to work successfully, it must understand its own constituency (for example, humankind) and convey this to the other side. It must also be capable of understanding the other's culture and goals and communicating these to its own constituency. As J. Stacey Adams has pointed out, people who serve in such roles are subjected to many cross-pressures.[30] From ET's point of view the human team may not be entirely trustworthy, because, after all, it will be representing humans and will put human needs first. At the same time many people may be distrustful of the human team because of its friendliness with the aliens. Conceivably, one of the human (or, for that matter, ET) negotiators could even switch allegiances and "go native." For many reasons, the negotiation team should be closely monitored and its members relieved from normal responsibilities and given strong psychological and social support.

Openness and Secrecy in Negotiations

There is an abundant literature, epitomized, perhaps, by Timothy Good's fascinating book *Above Top Secret*, suggesting that UFOs are alien spacecraft, that many of the world's governments have incontrovertible evidence of this, and that some governments may be in regular communication, perhaps face-to-face communication, with aliens.[31] This evidence, assembled by many UFOlogists, includes sworn testimony from unimpeachable witnesses; extensive photographic evidence, including thousands of feet of videotape or film; "trace evidence" in the form of chemical residues and physical changes at UFO landing sites; parts of the UFOs themselves (ranging from small pieces of wreckage to fully operational spacecraft); alien cadavers; and live ETs. In some of the more recent and breathtaking accounts, mammoth alien bases are supposedly sequestered underneath certain U.S. military reserves (in Nevada) where aliens and humans work side by side.

Yet, we hear, governments do not choose to acknowledge any of this. Instead, government representatives travel the country confiscating photographs and physical evidence of UFO visitations, quashing newspaper stories, and intimidating or silencing witnesses. Since, in the case of widespread sightings, mere suppression of evidence becomes increasingly difficult to do, a supplementary governmental strategy is mixing disinformation with fact, the former casting doubts on, if not totally discrediting, the latter. Curtis Peebles points out that, among UFOlogists, the issue of a government cover-up may have eclipsed the issue of the "reality" of UFOs themselves.[32]

A government that is open in its negotiations is in touch with its people and receives useful ideas and suggestions from them. Furthermore, it builds a sense of trust. Why, therefore, might a government wish to suppress evidence of contact with an extraterrestrial civilization, or keep the negotiations secret? Here, we will draw on the views of Allen Tough, who is not a UFO cover-up theorist but who is concerned that an urge for secrecy could make it difficult to formulate and implement an effective international protocol.[33] Tough's view is that the arguments for secrecy are outweighed by the arguments against secrecy, at least under the more probable contact scenarios.

One reason for secrecy is the threat of embarrassment. A government may suppress evidence of extraterrestrial intelligence because of the possibility that it is responding to a false alarm or is the victim of a hoax. To avoid the risk of humiliation it may err in the direction of suppressing the news. Tough counters that, with careful checking, the risk of making a mistake should be small and that it would be even more embarrassing to miss one of the greatest discoveries of all time.

Second, governments may be afraid to acknowledge contact lest they cause mass hysteria. This is the kind of reaction that one might predict on the basis of some people's reactions to Orson Welles's *War of the Worlds* broadcast. However, as noted in chapter 9, the public's reaction to the interception of an alien radio transmission is likely to be a far cry from public reaction to *The War of the Worlds*. Tough argues that the alien civilization is likely to be friendly and that there are ways to formulate and release the news to encourage calm and reasoned acceptance.

A third reason for secrecy is that news of an alien civilization will disrupt our cultural and social institutions. As we shall see in the next chapter, culture shock could threaten terrestrial religions, economies, and political systems. Tough believes that the unveiling of the extraterrestrial culture is likely to be slow, and that established human institutions are unlikely to fold at the "first call" from somewhere else.

Related to this is the prominent theme in UFO literature that governments suppress evidence of extraterrestrial intelligence because to do otherwise would result in their losing power. One of the main functions of the political and military establishments is to keep society safe against external threat. This could be difficult to do if the potential aggressor had (by terrestrial standards) an unlimited resource base and a military backed by a technology far in advance of our own. How much confidence would one have in an air force that had no choice but to allow alien spacecraft to roam one's skies at will? The gist is that by trying to maintain secrecy, a government prevents a loss of public confidence that could be so great as to result in its overthrow or at least defeat at the polls.

Fourth, Tough notes, a government may be tempted to maintain secrecy in order to gain an advantage over other nations on Earth. The first nation that intercepts a signal may choose to keep this secret until it has had the chance to decode, interpret, and perhaps even respond to the signal, thereby achieving a scientific first and a major propaganda victory. Then, too, a government may anticipate that the alien broadcasts will contain information that will confer a technological edge. By keeping this information secret, they could become more competitive in their industrial and military endeavors. Moreover, a nation may try to keep everything secret until it can propagandize the aliens or present its individual case in such a way as to gain favored status with the aliens. (As already noted, there are moral, legal, and practical justifications for ensuring that all of humanity is involved in human–alien contact, not just one nation-state. Besides, one hopes that a civilization many centuries beyond ours will be wise enough to avoid playing us off against one another.)

In sum, there are many reasons that the first nation or nations that make contact with an extraterrestrial civilization might want to maintain secrecy. According to Tough's analysis, these justifications either do not withstand close scrutiny (because they are based on superficial analyses or on improbable contact scenarios) or else they pale into insignificance in light of the justifications for a coordinated international or, better yet, supranational response. As psychiatrist Carl Jung noted: "Nothing helps rumors and panic more than ignorance. It is self-evident that the public ought to be told the truth, because, ultimately, it will nevertheless come to the light of day."[34]

Cover-Up Theories

Is the U.S. government suppressing evidence that President John F. Kennedy was not killed by a single psychopath acting on his own, but by a team of assassins controlled by a powerful crime syndicate or a foreign power? Is it true that the U.S. government is refusing to admit that American soldiers, sailors, and airmen listed as missing in action (MIA) in Vietnam and Korea are even now, 20 to 40 years later, still living in captivity? Is the U.S. government suppressing the fact that an alien spacecraft crashed in Roswell, New Mexico, in 1947? That alien cadavers are kept under continuous guard at Hangar 17, Wright–Patterson Air Force Base? (Or is it Lackland or Vance?) Could it be that SETI itself is a government-initiated sham, intended to draw our attention away from the fact that we have been negotiating with aliens for over 50 years?

Or what about the situation described by an informant to folklorist Thomas Bullard?[35] Briefly, a group of humans—including former presidents of the United States—has agreed to allow aliens to abduct humans in return for delivering certain advanced technology. The aliens have abducted far more humans than authorized, and they have transformed a small remote base into a series of interconnected bases spread out beneath several western states. At one point the government tried military action to expel the invaders, but casualties were too high. President Reagan's "Star Wars" proposal for advanced military technology was intended not to fight the Russians but to expunge the enemy from within. Meanwhile, many missing persons, including those children whose pictures we see on milk cartons, are held captive in underground bases, where their blood provides our unwelcome guests with essential nutrients.

Conspiracy or "cover-up" theories sell magazine and books, feed upon and reinforce our distrust of the government—and provide us with a measure of hope: that something momentous rather than minor caused the death of our president, that some of our long-lost loved ones might still be alive as prisoners of war, or that we are not alone in the universe. Cover-up theories help maintain beliefs in the reality of UFOs. One of the main problems confronting people who believe that extraterrestrial visitations have already taken place is explaining why there is so little proof of their existence. Government suppression of the evidence provides an answer to this persistent and annoying question. Investigators who attempt to expose cover-up theories run into the same sorts of problems encountered by those who study flying saucers themselves: tantalizing leads, evaporating evidence, long convoluted paths to nowhere.[36]

Claims in support of cover-up theories fall into several categories: behavior on the part of governmental officials and agencies that is suggestive of duplicity; purported government documents suggesting that the government is more knowledgeable than it cares to admit; eyewitness accounts that conflict with those of government officials; and accounts of government agents misleading people or openly intimidating them. Arrayed against this are allegations that the cover-up theorists are mistaken, misguided, or fraudulent. A critical analysis of allegations (and counter-allegations) requires expertise in history, document analysis, eyewitness testimony, character analysis, political science, sociology, organizational behavior, and psychiatry.

Perhaps certain features of organizational functioning tend to perpetuate conspiracy theories, whether they are about UFOs, MIAs, or the Kennedy assassination. On what kinds of bases might we infer duplicity or cover-up? Certainly, standing mute is suggestive of guilt, a fact well known to defense lawyers who would otherwise not put certain clients on the stand. Tardiness, or allowing an inordinate amount of time to pass before answering a question or releasing a document, hints of stonewalling, collusion, or censoring. The loss of a document or a piece of evidence is certainly suspicious, but even more self-incriminating are officials whose stories contradict one another's. After all, listening for inconsistencies is one of the ways that cops catch crooks.

Yet organizational inefficiencies and blundering can also foster each of these symptoms of deceit.[37] Large organizations are slow and make mistakes. Documents and evidence are lost. Spokespeople make statements based on partial and imperfect information, which are later contradicted by people who are more in the know. Add high emotions and time pressures, and the situation gets worse.

In the case of stunning events like the Kennedy assassination or UFO sightings, the evidence is distributed across many organizations. The files themselves are immense, measured not in pages or file drawers but in cubic yards. The information within them has been collected by a succession of investigators over a period of decades. This information is of uneven quality and in some cases contradictory. Files released under the Freedom of Information Act are likely to be highly censored. Of course, the agencies may be deleting the names of confidential informants or references to security matters that have nothing to do with visitors from outer space, but there is no way to know this. Given that the files are large enough to provide evidence for almost any point of view, that the censored parts of the documents could be the ones that prove one's theory, and that the smallest detail might be the crucial clue, is it surprising that cover-up theories die hard?

On close examination, though, the evidence can evaporate. In the late 1980s investigative reporter Dan Moldea demanded a reopening of the Robert F. Kennedy assassination. Moldea, like many other Americans, believed that assassin Sirhan Sirhan had not acted alone. Seemingly in support of this were many inconsistencies in eyewitness accounts, coupled with lost records and baffling evidence. Moldea found, however, that simple human error explained the suspicious circumstances. For example, bullets fired from the gun in 1975 had different markings from the bullets fired in 1968. Had the gun been switched? No, the rifling had changed because during the intervening years Sirhan's gun had been fired many times by detectives who wanted spent cartridges as souvenirs.[38]

In the late 1940s the U.S. Air Force was concerned about intruders in U.S. airspace, and at various points other agencies (including the CIA) were involved.[39] Within the U.S. government, some experts did support the "extraterrestrial" hypothesis: that UFOs were alien spacecraft. But as the years passed, commission after commission failed to find definitive support for this hypothesis, and the air force concluded that any threat that might exist was in the form of mass hysteria or perhaps mass confusion caused by false UFO reports at the onset of a Soviet missile attack. By the late 1960s, the government seemed to have lost interest in the issue and stood mute. Its refusals to discuss UFOs strike some as further evidence of a cover-up. As a psychologist, I suspect that by refusing to continue dialogue on the topic the government hoped that its harassers would eventually get bored and go away.

CONCLUSION

Although microwave observation is a passive search for extraterrestrial intelligence, once aliens are detected it is unlikely we will be able to remain silent for long. The problems of establishing diplomatic ties will be severe because of communications difficulties and a lack of understanding of their culture. Our response must be carefully thought out and made in behalf of all humankind.

Under some scenarios, at least, issues of jurisdiction will have to be solved. Because institutions will affect the course of interstellar negotiations, we must thoroughly understand the institutions that will be involved and try to overcome such traditional liabilities as a slow pace, a tendency to act on biased information, and an inability to identify a wide range of responses. For several reasons negotiation strategies based on threat and force are unlikely to work for us; for reasons discussed in chapter 7 it is unlikely that ET will favor coercive strategies either. Economic strategies, based on the exchange of goods or information, and integrative strategies, based on values and friendship, should have a better chance. Integrative strategies are particularly promising because all they require for their application is communication, which can be achieved by radio.

Although there are many reasons for negotiations to proceed in secret, there are more and stronger reasons to keep the public abreast of developments. Under almost any conditions some people will cry "cover up," but the evidence in support of cover-up theories tends to be flimsy, to evaporate over time, or to result from the normal inefficiencies that occur in organizational settings.

ELEVEN

THE ROCKY ROAD TO UTOPIA

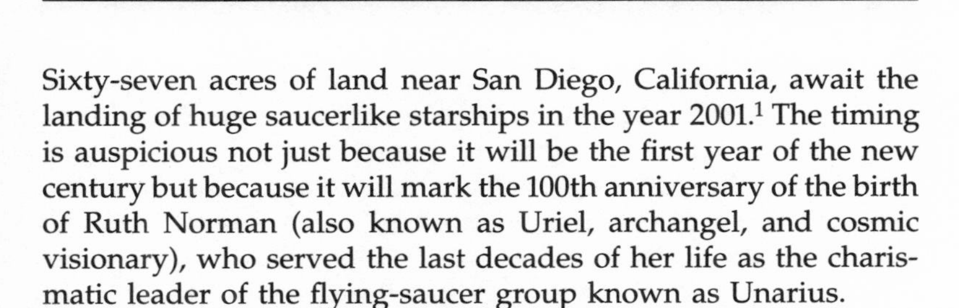

Sixty-seven acres of land near San Diego, California, await the landing of huge saucerlike starships in the year 2001.[1] The timing is auspicious not just because it will be the first year of the new century but because it will mark the 100th anniversary of the birth of Ruth Norman (also known as Uriel, archangel, and cosmic visionary), who served the last decades of her life as the charismatic leader of the flying-saucer group known as Unarius.

The fleet of "vehicles of life," each as large as 5 miles in diameter, will bring 33,000 scientists to Earth who will work for the benefit of humankind, bring us higher knowledge and new technology, and issue in an era of logic and reason. These space brothers will offer us "celestial alloys, extraterrestrial construction tools, and interstellar electronics." Their wonders will include caps that will implant new knowledge; new generations of computers grown from crystals; and "power towers" that produce cheap, nonpolluting energy. They will heal us, and they will guarantee peace. Because the arrival of the starships will solve all of our economic and technological problems, those of us who are not already doing so will be able to dedicate our lives to education and public service.

Whereas SETI enthusiasts may be loath to accept Uriel's prophecy that starships will arrive in the very near future, many do share an expectation that the first extraterrestrial society we contact will have abundant resources, amazing technology, unlimited material comforts, good health, few social ills, and countless opportunities to grow psychologically. In our mind's eye, they will have solved many of the scientific, technological, and social problems that confront us now, and will have reached a state of perfection that staggers our imagination.[2]

Given the opportunity, we should learn much from such a society. By their direct material aid or, more likely, by drawing on their knowledge, we could improve our technology and solve our medical, psychological, and social problems. Of course, such wonderful results would depend on their being able and willing to help us. These assumptions require close scrutiny, as do beliefs that the immediate solution of our most pressing problems would propel us toward utopia. Technological and social innovations have costs as well as benefits, and there is a risk that the intervention of even a benevolent alien society will create more problems than it solves. At the very least we shall have to proceed with our eyes open and recognize that we may get something other than we want. Rather than obey the adage "Never look a gift horse in the mouth," we may need to look very, very closely.

One analog for forecasting the results of an encounter between cultures of differing technology is the impact of major European powers (England, France, Holland, Portugal) on what are now regarded as members of the third world: China, India, and Africa.[3] Whereas each of these third-world areas had its own philosophical, artistic, and technical achievements, 17th-century European explorers and colonizers considered these societies scientifically and technologically backward, a perception that became increasingly pronounced over the next two centuries, as the Europeans themselves became more technocratic. During the 18th and 19th centuries Europeans felt compelled to undertake civilizing missions, which included forcing Africans and Asians to adopt such marvels as clocks, railroad trains, and the telegraph. The introduction of these devices was not invariably of benefit to the natives and in some cases had unintended sociopolitical effects. Nor were the consequences of this technology transfer entirely positive for the civilizing powers.

More instruction is found in the arrival of the English, French, and Spanish in the New World. Europeans brought horses, donkeys, and other amazing life-forms (including very tiny ones that caused diseases) and introduced stunning technology, including cannon, armor, swords, metal tools, mirrors, and a fabulous array of decorative items.[4] The arrival of the Europeans and their technology had drastic effects on many indigenous cultures. On the one hand, European tools eased labor and, in certain ways, improved the quality of the Indian's life. On the other hand, European goods created a wave of consumerism among the Indians that could not be fully supported even by their successful fur-trapping industry. In order to own European artifacts many tribes went into debt, and to pay their debts they had to sell their lands. Europeans also brought Christianity and were not necessarily understanding when American Indians seemed reluctant to embrace this new religion.[5]

Europeans themselves reveled in their technological superiority. Although at one time they had been somewhat respectful of at least some non-European people, they "began to confuse levels of technology with levels of culture in general, and finally with biological capacity." Their technological superiority and easy conquests warped the judgment of scientist and layperson alike.[6]

A different kind of precedent has been identified by Steven J. Dick, who argues that the standard anthropological models are inappropriate for forecasting the results of alien—human contact. European ventures involved people and artifacts arriving in a new world and then fanning out across the continent.[7] Human–alien contact is more likely to come through radio waves; consequently, we need prototypes where societies have been altered by ideas rather than by people and things. Such prototypes include the Arabs' transmission of Greek science to the Latin west in the 12th and 13th centuries; the introduction of Copernican theory during the 16th century, followed by the "galactocentric revolution" of the early 20th century; and reactions to the Darwinian theory of evolution.

Dick notes that Copernican theory gave birth to a new science and caused dislocations in theology. Darwinian theory, of course, has had a profound effect on both science and religion. Drawing parallels between reactions to these profound ideas and discussions following the discovery of extraterrestrial intelligence, Dick expects there will be, once again, strong immediate reactions, heated controversies, peaks of scientific research interspersed with periods of neglect, and above all, "the transformation of the way that we view our place in nature."[8]

Certainly we should appreciate Dick's point that the Age of Exploration provides an imperfect model for human–alien contact. Yet his arguments seem to reflect an assumption that the information that might be exchanged will be of interest primarily to scientists. If, in fact, their transmissions address political ideology, technology, popular culture, and the like, then the sociology of science may be an incomplete model.

Many universities today have ethnic studies programs dedicated to the arts, literature, culture, history, and sociology of various ethnic groups. One of the aspirations of such programs is to preserve such cultural elements. Complete acculturation or assimilation into the mainstream is unwelcome because the members of these groups consider it important to maintain a sense of identity. The threat is that when one culture takes over another, tradition and history are lost.

HELPING POTENTIAL

The optimistic, or "pronoid," view that our postcontact society will undergo rapid and positive development rests, in part, on what Frank White refers to as "the assumption of extreme results," that is, an assumption that contact will be the most wonderful thing that has ever happened to humanity.[9] In fact, notes White, the assumption of extreme results (which of course could also mean terrible results) is not necessarily justified. As Ben Finney points out: "Such extreme pronouncements exaggerate the probable speed and magnitude of the impact of radio contact.... Comprehending a totally different civilization light-years away, and absorbing the meaning of whatever messages were sent, would be a slow and tedious process calling for the efforts of specialists from many disciplines as well as the SETI scientists now engaged in the search."[10]

Variables affecting the impact of their ideas on us may be clustered under three headings. The first is their capacity to help us: the extent and pertinence of their knowledge and their ability to share this knowledge with us. The second is their willingness to help. The third is our receptivity to what they have to offer.

Capacity

The first requirement is that they must have information or other resources that humans can use: secrets of agricultural production, vaccinations against immunodeficiencies, cures for cancer and other diseases, principles to ensure international harmony, and so forth. One assumption is that all societies have to solve certain kinds of problems in the material, psychological, and social realms. As an older, more mature society, they will already possess solutions (or at least large parts of the solutions) to our problems.

Yet, as already noted, societies may develop in an uneven fashion across different areas or domains. Despite their technological supremacy, the Europeans of 1700 were trying to discover how to make porcelain of the same high quality achieved in Japan and China, and in 1870 British manufacturers still could not match the quality of a very thin opaque paper from East Asia.[11] Furthermore, their areas of maximum achievement may not correspond to our areas of maximum need. Because they live in the same physical world as we do, they should be well versed in such fields as mathematics, physics, and engineering. Because they are different life-forms, their medical, psychological, and social problems could be quite different from ours, and their solutions to such problems may not work for us. Consider, for example, the utility for us of immunological principles developed for silicon life-forms! Then there are the limitations imposed by diminishing returns. There are some areas where even thousands of years of additional tinkering yield few improvements. When an artifact (such as a toothpick) or an idea (such as democracy) is first introduced, there are likely to be rapid improvements. As it gets better and better, making it even better becomes progressively difficult. Variations become increasingly subtle, and in some areas we ourselves may be close to perfection.

We should not expect that every extraterrestrial society will have a sufficient resource base to assist every civilization that happens to come along. They could be knowledgeable and sympathetic, but so engrossed in their own problems, or so overburdened with requests for help (something like multimillionaires who are besieged with requests from charities) that they simply lack the resources to help.

Then, too, the extraterrestrials may not be able to deliver their information. Their possession of all the knowledge in the universe will do us little good in the absence of rapid interactive communication involving large amounts of relevant information that we understand and can use.

Motive

Simply because an extraterrestrial society is able to help us does not mean it will. Certainly, nations on Earth have, on more than one occasion, turned their backs on another's suffering. Just prior to the disintegration of the Soviet Union, the United States was very much concerned about transferring to the Soviets high technology that could have military applications. Why should a society from another solar system be any less guarded?

"Altruism" is a frequently posited motive for selfless action, doing something of benefit for someone else without the expectation of personal gain. In psychology, the concept of altruism has been supplanted by the concept of "helping behavior," because it's not possible to eliminate such motives as a hope for a reciprocal favor, a place in heaven, or some other future reward.

Within the animal kingdom, helping behavior is widespread. Parents in many species make great sacrifices, including the sacrifice of their lives, to help their offspring. The prevailing view among animal behaviorists is that this serves a useful evolutionary function. Whereas an individual may forgo personal survival (in the sense of further reproductive success) by its self-sacrifice for an offspring, it simultaneously increases the likelihood that at least some of its genes will remain within the gene pool. The animal mother that dies fighting off a predator or the uncle who leaves his fortune to his niece do not increase the odds of their personal survival, but they increase the likelihood that the family's genes will be perpetuated.[12]

For us to benefit from the extraterrestrials' help, their aid must cut across species lines. There *are* cases of cross-species helping behaviors, such as naturalists attempting to improve the welfare of animals, or a dog defending a human. But most cases of helping involve members of the same species rather than "outsiders." From an evolutionary point of view, it is easy to explain why members of the same species help each other, but difficult to explain why members of one species would help members of another, potentially competitive species.

More likely, any willingness of the ETs to help will reflect their values and a weighing of the potential benefits of helping against the potential costs. On the hopeful side, advanced societies may have achieved a high state of intergroup harmony. They could have an "internationalistic" perspective and be motivated to bring new worlds into their supranational system. Or, there may be less positive motives for offering aid. They could have a strong proselytizing streak and force other societies to adopt their technology and value system. (Chauvinism, of course, was one of the elements underlying the "civilizing missions" dispatched by Europe to the third world during the 19th century.) Or, they may justify their aid as "enlightened self-interest"—a desire to ensure that civilizations such as ours do well enough that we are unlikely to cause them trouble in the future. In a sense they could provide us with just enough aid to "keep the lid on our garbage can." Another reason for giving something away is to keep the recipient from taking it!

Of course, there could be motives for *withholding* help. Perhaps they will have developed a policy of noninterference based on a doctrine of letting sleeping dogs lie. Another deterrent might be a desire to leave a new culture uninfluenced so that it can remain pure for the purposes of study. Then there is the practical reservation that, in the long run, a culture such as ours might fare best if it is allowed to evolve on its own. Perhaps an orderly progress works better than the great leaps forward made possible by sudden access to the aggregated knowledge of an older and wiser society. Generally, when we are forced to think through the answer to a problem, we understand it much better than if we are handed the answer. John Barrow, for example, wonders if leapfrogging the scientific and cultural progression would sap our motivation, keep fundamental discoveries forever out of reach, and put us in the dangerous position of manipulating things that we do not understand.[13] A less pessimistic view is that ET's science and culture will be unveiled slowly, that the information we have about their world will forever be incomplete, and that learning from ET will engage us intellectually and spur our quest for knowledge.

One more possibility is that the delivery of direct aid, even though it costs little, works against the development of a balanced interstellar relationship. Among humans, a situation in which each side gives something and receives something in return is much more comfortable than a situation in which one side gives and the other side receives. All too often, the side that always gives eventually feels imposed upon, and the side that always receives feels impotent and guilty (feelings that are likely to turn to anger). Reciprocity need not always be immediate, nor must it invariably be of equal magnitude, but truly nonreciprocal relationships tend to fail.

Finally, the more powerful society that aids a less powerful one through the transfer of knowledge or technology loses some of its own power. Through technology transfer it grants the weaker society independence and the wherewithal to resist the more powerful society's influence. A simple example is the empire that arms its colonials, who then use the new weapons to stage a revolution.

Michael Adas points out that in centuries past African and Asian cultures provided Europe with an abundance of inexpensive raw materials that European technology could turn into valuable finished work. These societies, in turn, became markets for European technology and manufactured goods. But slowly, as the technology infused Africa and Asia, societies on these continents became increasingly self-sufficient and less dependent on European know-how. Over time, with the advent of western machinery and manufacturing processes, the "natives" grew capable of producing goods on their own and Europe lost them as customers. Some nations with relatively cheap labor became so efficient at production that their products began infiltrating European markets.[14]

Certain forces slowed the transfer of technology from Europe to Africa and Asia. Europeans retained ownership of mills, railroads, and telegraph companies. Furthermore, Europeans or only the most carefully chosen natives were allowed to run locomotives or operate telegraph keys. Similarly, whereas Europeans were willing to train Africans and Indians, this training was usually limited to the vocational school level and did not include the full range of topics available in the European universities. In this manner, by limiting and controlling the distribution of knowledge and technology, the Europeans prolonged their control over the third-world countries.

Given a decision to offer Earth aid, there remains the issue of delivery. Should information be dribbled out over time, with the intent of facilitating its absorption? Should information be shared in all areas, or only in areas that are the most pressing, that is, where aid will do the most good? Should aid be immediate and direct, or should it consist of equipping the recipient society to better help itself? To provide the best possible aid for a society where hunger is chronic, should a potential donor give food, or empower the society for increased agricultural production? The benefactor that provides direct aid in the form of foodstuffs may be obliged to provide it the next year and the year after that; otherwise hunger may return. If the means for production are offered and accepted, then the problem is solved once and for all.

Human Receptivity

A helping relationship involves both a giver and a receiver. There are many reasons the receiving culture might be unable or unwilling to accept the giver's largess. Our skepticism of the ideas of fellow humans has slowed our progress on many occasions. For example, the development of iron boats was delayed because of beliefs that heavy seas would pop the rivets, the sailors would bake alive when the boat sailed into the tropics, and that it would attract lightning bolts.[15] Why should we be any less skeptical of plans for something like a starship? After all, from what we know, it is impossible to exceed the speed of light, and collision with even a tiny object would destroy both starship and crew.

New technology will be particularly frightening if we suspect that those who propose it have evil intentions. One rather unnerving possibility, discussed by Allen Tough, is that we might be offered a Trojan horse, that is, something that looks good but that will in fact lead to our destruction.[16] A possibility here would be an interesting, computer-controlled biochemical or genetic experiment alleged to prolong human life but which introduces disease or creates dangerous new life-forms on our planet. Tough suggests that when we receive instructions for building or creating something we can undertake the necessary security precautions at that time. This presumes, however, that we will be able to understand the full implications of step-by-step instructions offered by a society that is far in advance of ours. As it does for most of us when we take prescription medicine, much will have to ride on faith.

Then, too, the extraterrestrials' aid may run counter to the values of our culture. Would we be willing to accept a gift of slaves? Would you eat livestock that could beat you at chess? Suppose we were given recommendations for gun control or abortion, or against the use of tobacco? Would "inside information" from a more advanced society resolve these issues once and for all? Or, would our preconceptions and belief systems make it nearly impossible to absorb and act on such information?

There is a rather anticlimactic reason for resisting aid: it may not be all that helpful. History is replete with episodes where forms of aid are offered that are of no particular interest or benefit to the receiving culture: coals to Newcastle, so to speak. Sometimes aid is based more on what the giving culture can provide or thinks that the receiving culture needs, rather than on the receiving culture's actual needs.[17] Western technologists sometimes spit out solutions to problems that don't exist. In the aftermath of a Guatemalan earthquake, relief agencies sent tons of food, even though the Guatemalan food production system had not been disrupted. The Guatemalan government tried to ban food imports, to no avail. The main result was economic disaster for Guatemalan farmers, stemming from an oversupply of food.

Arnold Pacey has chronicled some of the effects of importing technology that is not really understood. As a result of a drought that began in the 1960s, approximately 150,000 hand pumps were installed in village wells in India. The new pumps sometimes failed less than a month after installation, and at any point in time perhaps two-thirds of them were out of order. Pacey suspects that one of the biggest contributing factors was that the villagers "probably had exaggerated expectations of the pumps as the products of an all-powerful alien technology, and did not see them as vulnerable bits of equipment needing care in use and protection from damage."[18]

Ben Finney points out that people sometimes emulate other cultures in superficial, ineffective ways. A prototype here is the New Guinea "cargo cults"—Stone Age tribes that first encountered westerners in the 20th century. Westerners brought cloth, tools, and countless other luxuries for their own use and to trade with the natives in exchange for local products or for labor. But the members of cargo cults tried another way to attract those goods. They built crude mock airplanes, hacked runways into the jungle, constructed bogus radio shacks with vine antennas, propped together fake wharves, and in other ways reproduced in the most superficial fashion some of the accoutrements of western technology—all with the vain hope that their props and rituals would draw airplanes and ships laden with valuable cargo. The western agents' painstaking explanations did little good.[19]

Of course, we could go astray in many other ways. One is to accept new ways that are inferior to the old. For many purposes, the European firearm known as the harquebus was inferior to the bow and arrow. It was expensive to construct and maintain and difficult to operate; moreover, once it had fired and missed, the explosion scared off the prey. The bow and arrow was cheap to make and easy to fix, it could be operated rapidly, and its near silence typically gave the inexpert marksman a second chance. Yet it was often the harquebus, not the bow and arrow, that was the object of desire.[20]

Moreover, there's always the prospect of a techno-disaster. If the information is incomplete or if we are not really sure what we are doing, we could do terrible damage to ourselves. As it is, people accidentally expose themselves to radiation, inadvertently shoot one another, crash airplanes, allow reactors to melt down, poison themselves, and erect buildings that collapse. There will be no end of possibilities, given exotic, advanced technology. In a sense, we could be like fifth graders trying to operate a nuclear attack sub. Oftentimes, attempts to automate advanced devices and to build in safeguards only increase the magnitude of the disaster.[21]

HUMAN CONSEQUENCES OF TECHNOLOGICAL CHANGE

Major new technologies typically have ramifications that extend far beyond the most immediate and obvious. Technological change can affect human values, beliefs, and institutions. Although we may think of a new technology as solving a particular problem, it may have additional, *unintended*, consequences. Who would have anticipated, for instance, that simple hand mirrors could be lethal? Centuries ago, some American Indian "dandies" appreciated European mirrors because they could use them to monitor their appearance and paint their faces without enlisting help. The mirrors also revealed the terrible pitting of the face that followed some epidemics. Mirrors, which gave inescapable evidence of the extent of the disfiguration, set the stage for depression and even suicide.[22]

In his provocative book *Technopoly*, Neil Postman points out that the invention of the telescope, which made it possible to explore the heavens, revealed a picture that was difficult to reconcile with religious views that were premised on the uniqueness of Earth.[23] The printing press allowed individuals and families to possess Bibles and to do for themselves what previously was done by priests and pastors. Basic advances in the physical and natural sciences did not by themselves revolutionize human thinking; their full impact awaited widespread dissemination of ideas through the printed word. More recently, television has transformed humanity. "After television," notes Postman, "the United States was not America plus television; television gave a new coloration to every political campaign, to every home, to every school, to every church, to every industry."[24] Information Technology (or IT), suggests Rustum Roy, has such a pervasive effect on people's lives that it has become a "full-blown religion with its theology, a powerful priesthood, rituals, icons, and a claim to exclusive truth."[25] According to Roy, IT is one of the world's most powerful and rapidly growing religions, and one that transcends national and cultural boundaries.

Sociologists have long known that introducing new technologies leads to dislocations in a society's cultural and sociopolitical spheres. Oftentimes, a new technology is put in place before people are properly prepared, with the result that it is not put to good use or it creates new problems. For instance, the past century has been marked by successive waves of farm implements that accomplish tasks that formerly had to be done by hand. Such devices can throw tens of thousands of farm laborers out of work. Over time, retraining these workers or developing new lines of work mitigate the problem, but at first there is a gap between the technological development and the psychological and social adaptations to it. We refer to the gap between technology and the behavior of its human users as a "cultural lag."

Today, notes George Bugliarello, the growing imbalance between scientific achievements and the ability of society to use them effectively has put science at a crossroads. Although science's potential is greater than ever, in the absence of a new compact with society, science will fail. Scientists must reconnect with the rest of society by addressing such questions as progress, accountability, and purpose. Society is doomed if science and technology, with their overarching power, do not have a strong moral sense.[26] In the 1920s sociologist William F. Ogburn argued that technology was progressing at a much faster pace than society or culture, that whereas technological change was logarithmic or exponential, cultural change was linear. As a result, the gap between technology and culture keeps increasing over time. More recent writers such as Alvin Toffler, author of *Future Shock*, have agreed.[27]

Roberto Pinotti has discussed the rush of technological change in the context of a SETI-like contact scenario.[28] "Culture shock," notes Pinotti, occurs when a person from one culture becomes exposed to another culture. This results in "a breakdown in communication, a misreading of reality, an inability to cope." Yet, notes Pinotti, this disorientation is mild compared to "future shocks ... product of the greatly accelerated rate of change in society, a result of the superimposition of a new culture on an old one." Of course, Toffler was referring to rapidly accelerating internal change, that is, our society building on itself. Contact with a far more advanced culture could increase the pace many times over. The present effects of future shock, "malaise, mass neurosis, irrationality, and free-floating violence," notes Pinotti, "are merely a foretaste of what may lie ahead."[29]

The abolition of illness and crime, the development of foolproof weight-control programs, longevity measured in thousands of years, the elimination of work and drudgery, technological gizmos to satisfy every interest, and the answers to almost every question you ever had—certainly, each of these has a direct and immediate upside. The drawbacks may be a little more difficult to find, but we can begin our hunt in demographics, economics, religion, and politics.

Demographic Consequences

Clearly we would welcome help in the prevention and cure of illness. Humankind's medical advances over the centuries have been impressive, and our life span has steadily increased. Nonetheless, each year, millions of people are removed from our midst by cardiovascular illnesses, cancer, and a host of other maladies. Certainly we might expect extraterrestrials to possess superior medical knowledge that could diminish if not eliminate pain and postpone death. Solutions to such problems would extend our life span, but we would still be subject to the natural processes of aging and physical deterioration. Despite occasional claims of greater longevity, the greatest documented life span for a human being is just over 120 years. Consequently, truly long-term survival depends on a "fountain of youth," some substance or procedure for eliminating or reversing the aging process. Already we are on the verge of learning how to stop or rewind the cell's "clock," raising the prospect that some people who are now alive may live to be 200 years of age, and it is the healthy and productive years that will be extended.[30]

In effect, our new acquaintances might deliver what evangelists would like to or now pretend to deliver—healing, and everlasting life! Think how wonderful the prospect would be for people with illnesses, especially fatal illnesses, and for those who are approaching the end of a normal life span. With many more productive years we will be able to take multiyear sabbaticals from work and still build immense funds for a lengthy and healthy retirement. Our increased life span may make us more responsible, perhaps increasing our sensitivity to the quality of our environment and strengthening our desire to preserve it. On the other hand, both our psyches and our societies reflect our finite time on Earth, so a dramatically increased life span (never mind everlasting life) could cause strains elsewhere in the system.

How would we deal, psychologically, with almost indefinite life? Frank Tipler has estimated that the human memory can accommodate about a thousand years of experience—what would happen to the memories of people tens of thousands of years old?[31] What about the generation gap? How could newcomers hope to succeed in the marketplace when pitted against others with thousands of years of experience and no interest in retirement? Trying to induce change? How could you shout down the voice of experience? Waiting for an inheritance? Forget it! Think you have trouble remembering the grandchildren's names now? Wait and see! Managed to postpone divorce so far? Good luck!

Of course, a bigger problem could be overpopulation. Somewhere between one-half and one-fifth of all the people who have ever lived are alive today. This means, among other things, that if we had had immortality from day one, we would have at least two to five times the population that we currently have: two to five times the demands on space, two to five times the demands on Earth's material and economic resources, and so forth. Thus, solving the mortality problem would exacerbate the population problem. Of course, there could be "fixes" to the population problem, such as birth control or migration to the stars, but we know that the former will not be acceptable to all segments of the population and that the latter will bring its own set of economic, technological, and social challenges.

Economic Consequences

"Economics" refers to the distribution and utilization of resources. If we are to adopt a truly utopian point of view, we can look forward to unlimited material resources, energy, and know-how. We hope for a well-clothed, well-fed, well-housed world where each person has every imaginable material luxury. Perhaps more imaginable than a world which provides everybody with everything is a world in which expensive resources become cheap and much of the work that is currently done by humans, including "think work," is taken over by intelligent machines. People who have invested in key areas will lose their fortunes as once scarce raw materials become plentiful and as expensive technologies are replaced by cheap and efficient procedures. Certainly the stock market could reel; consider what would happen to the remaining drug companies if one company managed to obtain the rights to alien pharmacology.

Major advances in almost any area could dislocate thousands or even millions of workers or turn essential or desirable lines of work into meaningless tasks. A century ago, a train carrying 50 tons of freight displaced 13,333 porters and got the job done at a much faster pace.[32] The introduction of cheap sources of power could effectively wipe out the petroleum or electrical industries or at least reduce their workforce requirements to a very small percentage of the current level. Significant reduction of crime or social illness could put many peace officers and social workers out of business. Medical advances would dislocate tens of thousands of highly paid workers in the health professions. New educational technologies, which would advance us beyond our current, 14th-century level, could severely thin the ranks of university professors, school teachers, librarians, and others who support the educational mission.

Science and technology, noted Carl Jung, would be the first to be "consigned to the scrap heap."[33] Consequently, scientists are among those that are particularly likely to be heavily affected by the discovery of an advanced civilization.[34] Scientists devote many years to training and then develop a career, often a highly rewarding career, around research to understand the laws of nature and improve the condition of humankind. Contact with extraterrestrials could short-circuit their careers if the extraterrestrials answered the riddles of science and told us how to cure the world's many ills. In effect, scientists' lives could be rendered meaningless because they would lose the opportunity to achieve their scientific goals. Could the scientists' plight be compared to that of the witch doctor whose livelihood is threatened by the arrival of a fully equipped hospital ship? Sometimes, folk medicine and modern medicine exist side by side. We could, in effect, have two sciences: one that is comprehensive and correct but beyond our comprehension, and another that is somewhat lacking, but cast in simple terms that are understandable and acceptable to us.

Despite the talk about living lives of ease, many people's jobs are important to them for psychological as well as economic reasons. What would happen to the new jobless, of whom perhaps many were very skilled and heavily invested in their occupations? Would they be given a free ride? Would they want one? How would people who were *not* economically displaced feel about all of this? In discussions of job dislocation brought about by technical advances it is easy to talk about retrofitting or retraining people, but in the wake of massive technological changes what might these new jobs be? Retraining for what? These problems will not be easy to solve.

In the workplace, people respond to pay, fringe benefits, praise, and other material and social consequences of their performance. But extrinsic factors are only one source of satisfaction. People receiving identical pay or social approval do not perform at the same level. Those who are particularly energetic and persistent at their work also respond to various intrinsic rewards, that is, satisfactions associated with doing the work itself. We sometimes pursue activities with great enthusiasm because they develop our skills, pique our enthusiasm, or make us feel good about the high levels of accomplishment that we receive.

All of this could be lost with the wholesale importation of new technology. People who take pride in the fact that their skills are critical for a certain line of work or task may discover that new machines easily attain the same ends. Ten, 20, even 40 years of sustained effort may go down the drain when the superior civilization does our work for us. It is very difficult to imagine sustaining motivation when somebody or something else can accomplish the same objectives with practically no investment whatsoever. Thus, the side effect of finding new ways to get things done is not simply a matter of economic dislocation; it is also a matter of finding something challenging and productive to do.

A brighter prospect may follow a period of temporary dislocation. As Stuart Kauffman points out, although new technologies eliminate many lines of work they also create new ones.[35] The automobile, he notes, put stables, blacksmith shops, saddlers, and buggy manufacturers out of business, but it opened doors for road builders, automobile manufacturers, the oil industry, motel chains, and fast-food shops. He adds that, in the really old days, we hunted and gathered; now, almost all of us make a living in ways that were not possible for early humans. Perhaps even as *Homo habilis* could not envision theoretical economists making a living by scratching obscure equations on whiteboards, we cannot envision the productive activities of our own distant descendants.

Political Consequences

Contact with extraterrestrial society could have major political implications. The discovery of extraterrestrial intelligence would affect both individual societies and international relations. Politicians who are in power could have difficulty maintaining this control if people believed that their government was powerless or unable to defend them against space invaders. One of the reasons alleged for the "cover-ups" of the 1940s and beyond is that the government did not care to reveal its total inability to prevent the flight of mysterious objects over the country. The worry is that the perceived helplessness of the government would lead to a total loss of confidence in it, accompanied, perhaps, by panic, riots, and (horror of horrors) incumbents being booted from office.

Pinotti argues that the superimposition of an alien culture on our own would result in a situation that we would find incomprehensible; in effect it would rob us of the rules that allow us to function in present-day society. This, he says, would lead to a "worldwide authority crisis." One possible response to such a crisis is magnified ethnocentrism, an attempt to preserve one's past and values by espousing them in an exaggerated form. This attempt to survive a confrontation with alien life would lead, says Pinotti, to the fragmentation of multicultural societies. Such fragmentation could create significant authority problems for federal and supranational governments.[36]

Contact could disrupt equilibrium within the supranational

system in many ways. Through some combination of effort and luck, they could make contact with an individual nation, rather than the United Nations or some sort of bloc, or the contact could be made with a bloc, but not of one's own choosing. Thus, one country, perhaps a relatively weak country, perhaps a "renegade" or third-world country, could gain a powerful ally, thereby upsetting the balance of power. Instead of being our "saviors," the aliens could form an alliance with our enemies. Of course, there may be ways to minimize such threats. Certainly, efforts to transform SETI into an international activity governed by strict protocols represent steps in that direction. Similarly, Donald Tarter's recommendation for an immediate, authoritative response that would preempt other responses is intended to minimize destabilization of the international system.[37] Nonetheless, a certain amount of political dislocation may result from contact, which could be particularly unpalatable to political leaders and groups that are currently well placed within the power structure.

Human Values and Spiritual Consequences

The rapid, wholesale importation of new technology can challenge the human spirit. In his book *Technopoly*, Neil Postman discusses the relationship between human technology and human institutions.[38] He identifies three historical epochs that are defined by the relationship of cultures to their technology. In the earliest epoch, that of tool-using cultures, religious beliefs, values, and social institutions provided the glue that held societies together. Traditional values and religions sustained a sense of history, perspective, and continuity. Beyond helping us to get work done, tools served important symbolic functions (for example, the construction of churches), but technology was secondary to ideology and human values prevailed.

The second historical epoch saw the rise of technocracies. At that time basic values, religions, customs, and traditions coexisted with technology, but technology itself became increasingly prominent, and, on occasion, the truce between them was a bit uneasy. Custom lost some of its hold, but people were still guided by traditional values.

The third and most recent epoch, dating back perhaps a hundred years, is the technopoly. This, our current era, is the era of "totalitarian technology." Traditions no longer define human purpose or give us an overall sense of perspective. Technology shapes religion, art, the family, politics, and history. It involves the "submission of all forms of cultural life to the sovereignty of technique and technology" and diminishes our social, political, historical, and metaphysical foundations.[39]

Postman also notes that, on some occasions, technology has undercut human values, but left nothing in its place. For example, during earlier times, the Bible and religion held great sway over almost all western societies. The Bible gave us both an accounting of our origins and a set of moral precepts. Theories such as Darwinism and scientific discoveries such as the telescope displaced creationism and the "anti-Copernican conceit" that our planet and solar system held a privileged place in the universe. The shift from creationism to science weakened our traditional moral precepts, which, one gains from Postman, was a form of discarding the baby with the bathwater.

Postman is neither a Bible-thumper nor some sort of latter-day Luddite intent on destroying technology; he recognizes that technological advances have positive as well as negative features. What concerns him is the mindless proliferation of technology, often for its own sake and without regard to its ultimate human costs. He is concerned that human values no longer control technology; instead, technology reigns in its own right and sometimes creates values that are devoid of human purpose. Our values and beliefs should control our technology, not the other way around.

Another worry about a possible alien–human contact is that by showing that human beings are not unique, contact will undermine human religions. Yet, as shown in Karl Guthke's discussion of the interrelationship of western religious views and speculation on extraterrestrial intelligence, a belief in life on other worlds is not necessarily heretical. The degree of incompatibility between religious views and the belief in life elsewhere has depended very much on time, location, and the manner in which the views were presented. Exochristology, says Guthke, is alive and well.[40]

Ted Peters notes that contemporary Christian theologians do not seem to have paid all that much attention to "exotheology—that is, speculation on the theological significance of extraterrestrial life." However, many of those who have explored the issue look favorably on the idea of many worlds and are not threatened by the prospects of extraterrestrial intelligence. These theologians include George van Noort, Theodore Hesburgh, Francis J. Connell, Krister Stendahl, and A. Durwood Foster. Peters quotes Billy Graham as telling an interviewer that "I firmly believe that there are intelligent beings like us far away in space who worship God.... But we would have nothing to fear from these people. Like us, they are God's creation."[41]

Michael Ashkenazi, of Israel's Ben Gurion University, conducted interviews with 21 theologians; 17 were virtually certain that extraterrestrial intelligence exists.[42] Ashkenazi concludes that many of the world's great religions, including Buddhism, Hinduism, and the Chinese religious complex, either explicitly acknowledge the possibility of advanced life-forms elsewhere or at least do not make a sharp distinction between the human and nonhuman. Somewhat more complex are the "Adamist" religions—Christianity, Judaism, and Islam—which believe that man is created in the image of God. Ashkenazi expects that the more hierarchically organized Christian religions will make a determination, and then members will fall in line. In the past the Catholic church concluded that American Indians had souls and were eligible for conversion and salvation; at some future time they may make the same determination about extraterrestrials. Protestant sects are relatively autonomous, and we might expect a variety of responses, with the most fundamentalist and dogmatic sects reacting the most adversely. A leader of a relatively small and unimportant sect could gain power, and do great damage, through combining television, fundamentalism, and fire and brimstone to whip up hatred against the aliens.

Variations among Christians are well illustrated in Howard Blum's recent discussion of differing reactions to a small midwestern town's decision to host a flying-saucer festival, much as other farming towns might hold an Asparagus Festival or host Cucumber Day.[43] One minister launched a diatribe against a "vague, misguided paganism," but other congregations were not upset. Some recognized that the flying-saucer festival was political and economic, rather than idolatrous. In one church, a small card appeared carrying parts of a quote from Father Theodore Hesburgh, a former president of Notre Dame who considered SETI an important vehicle for knowing and understanding God.[44]

Still other intriguing postcontact issues could arise once we learn about ET's religious beliefs. For example, if they lack such beliefs, they might be seen as ripe for conversion by those of us who belong to religions such as Islam and Christianity that are intent on saving others. However, Ashkenazi also notes that simply because the aliens' society is highly advanced will not mean they eschew religion. We may be affected by significant similarities to, or differences from, terrestrial religions. Of particular interest would be beliefs or practices that conflict with our own: for example, necrophagy, or eating one's dead.

Jacques Vallee proposes that an understanding of ET's religious beliefs will help us evaluate our own existential views. He has cataloged many of the existential positions advocated by different religions and notes that it will be interesting to see which ones are advocated by extraterrestrial civilizations. In addition, some of the views he has cataloged are capable of proof, and depending on what we find, some terrestrial views may gain ascendance and others retreat into insignificance.[45]

CONCLUSION

If advanced extraterrestrial societies exist, if we find them, and if we understand them, then there may be ways to benefit from their experience and knowledge. Although the difficulties of interstellar travel mean that they are unlikely to arrive in person, bearing gifts, we can still benefit (or suffer) from their ideas. Whereas many of these will be of interest to scientists, the rest of us may also gain exciting new technology and its practical applications. Ultimately, these could change not only the way that people view the world but also their daily lives.

We cannot simply incorporate extraterrestrial ideas without thinking them through, because our systems (supranational, societal, and organismic) have highly interrelated parts, so that changes in one arena yield changes in another. Although it is tempting to throw the spotlight on individual problem areas where a richer or wiser civilization could offer us help, many problems are very much related to one another. Dramatic progress in one area may exacerbate problems in another unless we take care to think through the implications for the system as a whole.

Although it will be divided, public opinion is likely to favor the rapid embrace of promising new technology. Its immediate benefits are likely to be much more obvious than its side effects or long-term consequences. (Smoking may be a useful analog: presumably, smokers are attracted to their vice at least in part by the immediate benefits of increased alertness, focus, and energy rather than by the long-term consequences of emphysema and lung cancer.) Although media coverage encourages us to expect extremes (in its "New Hope" or "No Hope" reporting), far more often than not press reports on technology are promotional and play on our desire for easy solutions to problems. The media's tendency will be to discourage critical thought to potential social and economic problems, such as worker displacement or the limited number of potential beneficiaries.[46]

The broader issue is, who is in charge? Should it be a benevolent extraterrestrial civilization whose suggested solutions may or may not be of value to us? Should it be the technology itself, which seems to have a logic of its own? Is it up to us to adapt to their technology? Discussions of cultural lag and related topics usually put rapidly advancing technology first and then bemoan people's inability to catch up. There is something wrong with the worker who is intimidated by a new computer, not the computer or its instruction manual. As Pacey has pointed out, the question should be put the other way around: Given human needs, how can technology catch up?[47] People, not things, should be the driving force.

Of course we should entertain new ideas, not only in science and technology but also in philosophy, religion, and the creative and performing arts. The point is that if we are to progress beyond the detection of carrier waves or the exchange of pleasantries, we need to be aware that the road to utopia may not be a smooth one. We should sit firmly in the driver's seat, and remember the warning "Be careful what you wish for, because you might get it."

TWELVE

BETTING WITH THE OPTIMISTS

As a child during the closing decades of the 19th century, Protestant theologian Paul Tillich lived in Schönfliess-Neumark. This medieval German town, built around a Gothic church and surrounded by a high wall, constituted a small, self-contained, but narrow and restricted world. Eventually, a railroad reached Schönfliess, enabling Tillich to travel elsewhere, including the Baltic coast and Berlin. Both the sea and the city offered him "infinity, openness, and unrestricted space" and generated exciting new possibilities. His life was never again the same, and his fascination with enlarged horizons affected both his choice of places to visit and his theology.[1]

In a sense, Tillich's childhood adventures give us an excellent metaphor for SETI. We live in a relatively small and self-contained world that is in some ways confining or restrictive. Despite our occasional forays into space, gravity forms an invisible wall that for now restricts almost all of us to the surface of our own planet.

Yet our probes, telescopes, and radio telescopes allow us to peer outward. In 1940, England's astronomer royal, Herbert S. Jones, acknowledged that we "cannot refrain from toying with the idea that somewhere in the universe there may be intelligent beings" who are our equals or perhaps our superiors and who have "managed their affairs better than we have managed ours." He added that the work of astronomers or biologists cannot help us resolve this matter and concluded that it must "remain forever a sealed book."[2] Scientific developments since Jones's time make his conclusion seem both premature and pessimistic. If there are other electromagnetically active civilizations in our galaxy, radio telescopes give us a chance to find them, and if we can interpret their transmissions, we may be able to pry open that "forever sealed" book. Many of us will be profoundly and permanently influenced by our enlarged horizons, and our lives will never again be the same.

It has been a circuitous journey from discussing the number of stars in our galaxy to considering the possible amelioration of Earth's problems with the aid of benevolent extraterrestrials. Our itinerary has emphasized the behavioral dimensions of the trip: how psychological and cultural factors shape people's attitudes toward SETI and define the search; what the plausible similarities and differences may be between human and extraterrestrial biological and social entities; and how alien–human contact may affect Earth's people, societies, and cultures. Despite the occasional reservation or doubt, a strong optimistic theme prevails in my assessment of the prospects for SETI, the nature of extraterrestrials, and our reactions if "they" are found.

In this book I have developed the theme that intelligent extraterrestrials exist; that we will eventually uncover incontrovertible evidence of their existence; that their life-forms and their societies will be recognizable and to some extent understandable to us. When contact occurs we are more likely to handle ourselves well than to panic, and whereas we cannot expect a magical solution to all of Earth's problems, we are more likely to gain than lose from exploring their ideas. We conclude our journey with a final look at the ideas that have guided us through certain junctures, and show why, when we come right down to it, we should place our bets on the side of the optimists.

PROSPECTS FOR SETI

Through recorded history people have wondered what life-forms and civilizations, if any, might be found among the stars. The weight of opinion on the existence and prevalence of extraterrestrial civilizations has swung back and forth depending on politics, religion, and the state of science. For centuries, evidence has grown that there is nothing unique about our sun or planet. During recent decades in particular, scientists have strengthened their suspicion that life, including intelligent life, has originated again and again. Given advancing technology and sufficient time, we will come to know some of these life-forms.

The Likelihood of Extraterrestrial Intelligence

The Drake equation (see chapter 1) is a helpful procedure for estimating the number of technologically advanced civilizations in the universe. The appeal of the Drake equation is its ability to organize, in very broad strokes, our thinking about the physical, biological, and social requirements for a long-lived civilization. The various assessments it asks us to make are not equally easy. In essence, Drake starts with simple questions that lead to more difficult ones. We are off to a great start by determining that there are a huge number of stars. Discoveries during the late 1990s increase our confidence that many of these stars have planets.

The guesswork becomes more difficult when we try to estimate the proportion of planets capable of hosting life, especially given that the planets we are capable of identifying right now are unlikely prospects. We know that the elements essential for our form of life are common in the universe. A little encouragement comes from theories that cosmic processes begin forming the most rudimentary genetic plans, and that microorganisms may survive interstellar travel. Even more encouraging are theories that the origin of life depends on self-organizing systems rather than on a random shuffling of molecules until they manage to achieve a pattern that overcomes all-but-impossible odds.

Perhaps the most critical juncture in the Drake equation is estimating the likelihood that life-forms will develop sufficient intelligence to begin microwave broadcasting. Optimists should be bothered that some of our best experts on life, paleontologists and evolutionary biologists, have significant reservations about the evolution of intelligence in any recognizable form. To be sure, certain anatomical structures that are important for information processing appear to have evolved "independently," and certain key evolutionary trends have never been reversed, at least in recent geological times. Some comfort might be found in the hypotheses that order underlies the chaos in the geologic record and that self-organizing processes move inert matter in the direction not only of life but also of consciousness.

The bottom line is that the ordered view of the universe proposed by the physicists does not square with the role of chance as envisioned by evolutionists. Are physicists and other SETI enthusiasts overly committed to an ordered view of the universe and naive in their "alien designs" or anthropomorphic projections? Or do evolutionists assign too much weight to chance and defend an incomplete theory? Paul Davies believes that the discovery of independently evolved life will be one of the greatest discoveries of all time because of its implications for our thinking about evolution.[3]

Many recent discoveries—such as new evidence of the existence of planets or at least planetlike objects, discoveries of possible traces of life on meteors from Mars, hints of water on distant moons—shift the odds in the direction of more and more stars surviving the successive rounds of elimination imposed by the Drake equation. Yet, the case for ET remains circumstantial: we have yet to prove that even one planet aside from Earth harbors intelligent life.

Balanced against mounting circumstantial evidence is the "great silence," the lack of acceptable proof that extraterrestrial life-forms exist. Some of those who have written in support of a pessimistic conclusion—we have not yet seen them because they do not exist—presume that interstellar travel is the preferred mode of interstellar contact. Pessimists suggest that although interstellar travel stretches the frontiers of our technology and may be enormously expensive and difficult, it may not be beyond the grasp of more advanced civilizations. Certainly, if the optimists are correct, some of these civilizations must be very old; they would have had ample time to reach us by now. Although psychological, social, and cultural factors may have dissuaded some of these ancient societies from undertaking interstellar migration, certainly no "sociological" explanation could apply to all such societies. It does not matter if 99 percent of all societies develop introversion, a fear of foreigners, or other characteristics that keep them at home. All it would take is one technologically advanced and aggressive society to expand throughout our galaxy.

Optimists counter that even if interstellar travel is possible it is not necessarily feasible. They suggest that we have not yet conducted the kind of search that would reveal evidence of ET's presence. Moreover, we should not categorically dismiss sociological explanations of the great silence. The case against sociological explanations is flawed because it fails to reckon with the dynamics among states. Expansionist tendencies are likely to be tempered by neighboring states, even as belligerent terrestrial societies are kept from running amok by their neighbors on Earth. It may be that all it takes is one technologically advanced and aggressive society to begin expanding throughout the universe, but all it takes is one or two other societies to hold those expansionist tendencies in check. When we raise our sights to include dynamic interstate systems, the viability of "sociological" explanations of the absence of evidence is restored.

Spirited debate on the great silence has many useful consequences for those of us who are interested in life in outer space. The debate forces optimists to review their arguments in ways that would not be necessary in the absence of opposing beliefs. Pessimists offer us thoughtful discussions of extraterrestrial strategies for interstellar travel (for example, the use of nanotechnology rather than huge, lumbering worldships) and search techniques (for example, the use of radar) that could reveal these strategies in action. Moreover, their discussions of the technologies, motives, and time tables for interstellar migration increase our confidence that, in the long term, we ourselves will become capable of migrating to the stars. In a sense, the pessimists become the optimists and the optimists become the pessimists when we shift the discussion from the prospects of finding extraterrestrial intelligence to the prospects of human conquest of space.

The Likelihood of Contact

Apart from mental telepathy, which does not meet acceptable standards of reliability, societies on different planets may encounter one another in two basic ways: travel and radio communication. Manned space travel (or its extraterrestrial equivalent) is expensive, difficult, and fraught with hazard. Robot probes are a possibility, particularly for a rich and patient society. Whatever the prospects for different strategies, almost all of the actual empirical searches involve using radio telescopes to search for signs of other electromagnetically active civilizations. The underlying assumption, and it is a big one, is that other technologically advanced civilizations are transmitting microwave signals of sufficient power as to be detected by us. Once we accept this assumption, there is much to commend a microwave search, including its relatively low cost, the ability to explore the possibility of life existing near any one of thousands or millions of stars, and the possibility, at least, of quick results. Of course, if all civilizations simply listened, then no two civilizations would ever make contact.

Necessarily, economic and technical factors constrain the search. They limit the number of places we can look, the range of frequencies we can explore, the minimum strength of the signal we can detect, and the amount of time we can spend. Some of us may question the choices that define the search space, or decry as anthropocentric the convictions that their scientists will think like our scientists and that radio traffic will gather around some sort of central gathering place or cosmic water hole. Yet there has to be some place to start, and over the years SETI technology has done nothing but improve. Thus, the search becomes increasingly thorough. It doesn't do any good to lament strategies beyond our grasp; it makes sense to forge ahead with ways accessible to us.

If search procedures confirm a carrier wave of extraterrestrial origin, there will be an attempt to determine what information, if any, it carries. The usual assumption is that their message will be built on mathematics and physics, since these must be common reference points for all civilizations that use radio. John Baird has questioned this assumption; he points out that when he communicates by telephone with a scientist friend one of the topics that never arises is the telephone itself.[4] However, as a former amateur radio operator, or "ham," I offer a counterexample. The hundreds of thousands of radio hams who live all over the world and from almost every walk of life find little to talk about on the air than their transmitters, receivers, antennas, and signal strength. If the goal itself is communication, it makes sense to discuss the characteristics of the communication link.

Deciphering the transmission, if possible at all, will be difficult and time consuming. Our experiences thus far on Earth are not entirely encouraging. We can "communicate" with only a handful of the hundreds of thousands of animal species that live on Earth, and we do this in such a restricted way that some people question whether we can really communicate with any of them at all. We have trouble deciphering ancient inscriptions such as Egyptian and Mayan hieroglyphics; it was only through the discovery of the Rosetta stone, which presented the same message in known languages and in Egyptian hieroglyphics, that we became able in the past 200 years to understand some of what the early Egyptians wrote. There will be no Rosetta stone, at least not at first, because any interstellar "Rosetta stone" could not include any languages known to us.

Whether we will be able to understand ET's transmission is anybody's guess. It may depend on whether we eavesdrop on an "internal" communication, intended for members of the same society or supranational system, or intercept a broadcast intended to attract members to the Galactic Club. Extraterrestrials are likely to have given great thought to constructing a message that beings who are very different from them can decipher. Unless they are just beginning their own quest for intelligent life in the Universe, they may be quite experienced.

UNDERSTANDING ET

One can take three basic positions when the discussion turns to the risky area of what they are like. The first position is to admit to the possibility of an infinite variety of life-forms everywhere, including those that may suspend themselves in frozen gas or lurk beneath planetary surfaces. The second is to assume that nature has limits and to attempt identify some of the general ways extraterrestrials might resemble living entities on Earth. The third position is that they will have evolved in the same way that we have evolved and will bear a remarkable resemblance to humans. As we progress from the first to the third of these positions we sacrifice abundance (the more unusual life-forms we can imagine, the greater the number of inhabited niches in the universe) but gain detail.

The second or intermediate position—that they are something like us—neither stretches us beyond our understanding of physical or biological principles nor forces us to rely on a common genetic plan or a miraculous degree of convergence in the evolution of terrestrial and extraterrestrial species. The intermediate position allows us to use our knowledge of biological and social entities on Earth to organize our thinking about intelligent life elsewhere in the universe. The intermediate position has guided this discussion of the possible nature of extraterrestrial intelligence. It has an optimistic ring, because it suggests that we can prepare ourselves to understand extraterrestrial life-forms and that when contact occurs we will not be completely befuddled.

We began with the assumptions of monism, empiricism, and determinism, and the conviction that the same basic substances and processes are found everywhere. We used, as the basis for extrapolation, our one known case, Earth. Although Earth is only one case, nested within it are a plenitude of examples—millions of species, thousands of cultures, hundreds of nations spanning a written history extending back perhaps 5,000 years, and physical evidence that extends back many times longer than that. Given encouragement to organize our thoughts about extraterrestrial intelligence, this provides a good place to start. Earth may be only one case, but it is one more case than would be available if we chose to ignore it.

James Grier Miller's Living Systems Theory provided the framework for our discussion of hypotheses about extraterrestrials.[5] This theory stresses the continuities among physical, biological, and behavioral sciences. For billions of years, life on Earth was dominated by blue-green algae. Change has been in the direction of increasing size, differentiation, integration, and hierarchical levels. Individual organisms evolved and banded into groups that became successively larger: communities, societies, and ultimately supranational systems. Each of these system levels represents another step forward in the organization of individual cells.

Living Systems Theory is in many ways compatible with the thinking of SETI scientists. It begins with the most fundamental units practical for exploring individual or collective behavior. These elements—matter, energy, and information—are already well understood by SETI scientists and shape their thinking about aliens and their civilizations. Miller's theory views living systems as open systems, and its analyses recognize structure, function, and parallel subsystems and processes. It seeks uniformities across species, cultures, and historical epochs. The principles that are seen in operation again and again on Earth are more promising places to start than are principles that have never been observed at all. Thus Living Systems Theory; data from anthropology, biology, political science, psychology and sociology; and the ideas of astronomers, exobiologists, and others who have had the temerity to speak out on the topic formed the basis for our discussion of the possible nature of ET, ET's society, and the larger social framework in which that society is embedded.

Views of Organisms

Although extraterrestrials could include some very exotic types of life-forms, scientists expect them to be carbon based. They will be animal-like in the sense that they will have shape and mass, as well as the ability to move from place to place, use tools, and affect their environment. They will be large enough to support high-capacity information-processing systems, to build fires, and to work metals. They will store and metabolize energy and will not require near-continuous feeding.

Their nervous systems will differ from ours, but they will be capable of detecting conditions in the external environment and of storing, retrieving, and utilizing information. They will have vision, or the equivalent of vision, which will allow them to observe the stars. They will be a lot smarter than we, and we will probably think of them as more successful. They will have language, a way of understanding math and physics, and a spirit of inquisitiveness. ET will not necessarily look like us, but if there are only so many ways to succeed in nature, we should not exclude that possibility. Perhaps we will communicate not with flesh-and-blood organisms but with a cyborg or with artificial intelligence. Or perhaps we will have it both ways if we discover an emulation of a living organism embodied within a computer. Depending on the form of their computer technology, this might be our best chance for encountering a silicon-based life-form.

Views of Societies

We will not discover an isolated individual but rather one or more members of a large, self-sufficient social system known as a society. Like terrestrial societies, extraterrestrial societies will have some form of culture that shapes their members' views and influences their choices and activities. Very useful for us is Bohannon's view of culture that, like Living System Theory, places strong emphasis on the continuity of the physical, biological, and social sciences. Everything flows from matter and energy; culture is not apart (separate) from nature; it is a part (integrated) with nature.[6]

Experience gained from observing all known terrestrial cultures and many nonhuman species suggests certain functional prerequisites for social living. We should expect some form of dominance hierarchy or social stratification, a division of labor, means for coordinating with one another, and a sense of responsibility for each other's welfare. Extraterrestrials could be situated on a planet, a satellite, or an interstellar worldship. Their society is likely to be old and, from our perspective, materially well-off.

In the course of solving matter–energy and information-processing problems, societies may encounter certain potentially fatal problems. Nuclear power could be used for destructive purposes or without proper safeguards. Beliefs and behaviors that are useful for increasing the population to a certain point—achieving satisfactory levels of societal staffing, so to speak—may continue long after there are no economic or political justifications for continued population growth. Cheap, efficient transportation and manufacturing processes may yield undesirable side effects such as pollution.

Societies that encounter such problems face three potential outcomes. Some will die off, and because they no longer exist we will never hear from them. Others will survive through moderation: rationing resources, prohibiting the use of certain technologies, and so forth. Such societies can't afford interstellar transmissions, so we won't hear from them, either. The other societies will work through their problems without depleting their resources or inhibiting economic growth or the development of new technologies. It is these successful societies that we would encounter. Should we be encouraged or discouraged by events on Earth? Adrian Berry's recent work suggests that developments over the past few centuries have been predominantly positive, that assessments of Earth's current problems tend to be overstated, and that there are bright prospects for the future.[7]

The ultimate problem of social organization is placing enough limits on individuals so that they can coordinate with one another and fulfill their basic responsibilities to the society while nonetheless leaving them sufficient latitude to satisfy their individual needs and experiment with creative solutions to problems. In the absence of sufficient restraint we may expect anarchy; in the absence of sufficient freedom we may expect rebellion. Centuries of trial and error on Earth are leading to a prevalence of democracies. If, as some political scientists suspect, the trend toward democracy is based on truly fundamental, immutable principles, the extraterrestrial form of government may be very recognizable to us.

Views of Supranational Systems

If many extraterrestrial societies exist, some must have already encountered one another. Thus, the concept of the Galactic Club. Whereas there are many possible interstellar arrangements, the trend on Earth points to supranational systems, groups of societies whose members have relinquished some of their sovereignty in return for coordination in one or more functional areas such as arms control, the preservation of environmental quality, and the regulation of transportation, communication, and commerce.

Although it makes sense to proceed with caution and attend first to our own security, chances are that we will not encounter an "evil empire." Their large resource base and advanced technology are unlikely to steer them toward war. At interstellar distances, communications and command and control problems are formidable for prospective galactic conquerors. If democracies prevail elsewhere as they prevail on Earth, we can take heart, because democracies do not go to war with one another, only with authoritarian states. Computer simulations show that nation-states that refrain from aggression but that do come to one another's defense form stable interstate systems, and that within such systems the more peaceful societies outlive the more aggressive ones. For centuries now, there has been growing recognition that war is morally repugnant and methodologically ineffective. War itself is becoming rare, at least among large, economically strong nations, and it is possible to foresee world peace. We aren't there yet, but the number of centuries that it has taken us to shift from warring clans to the United Nations would be but a blink of the eye in the history of a civilization a billion years old.

Thus, a fundamentally positive picture emerges when we extrapolate from life on Earth: there are trends toward democracies, the end of war, and the evolution of supranational systems that impose order on individual nation-states. This suggests that our newfound neighbors will be peaceful, and this should affect our decision about how to respond to them.

Hedging Our Bets

Although Living Systems Theory focuses on commonalties, it does not obscure differences. Despite common foundations in matter, energy, and information, despite common subsystems, structures, and processes, the living systems on Earth differ tremendously. Despite a common planet and a common genetic heritage there are tremendous differences among people. Although we may have some general expectations, we will not know the extraterrestrials in their specifics, and we will do well to remember that our own ways of viewing the world may be very different from theirs. Whatever the ultimate merits of Living Systems Theory, whatever the extent that we can generalize from using ourselves and our Earth as starting points, we will have to proceed with caution. Rejecting the view that imputing non-humans with humanlike characteristics is invariably wrong is not the same as accepting the view that they will be just like us.

Hypotheses about extraterrestrial intelligence are hypotheses, not facts. A fine line separates the rational process of extrapolating our knowledge of life on Earth to life elsewhere and the irrational process of projecting fantasies, wishes, or fears onto unknown entities whose very existence is in doubt. We try to do the former without lapsing into the latter, but given our present state of knowledge, the question is not whether but how often we slip across the line.

CONTACT AND ITS AFTERMATH

The discussion of contact—confirmed evidence of the existence of intelligent extraterrestrials—and its aftermath moves us back to slightly firmer ground, for whereas we can only speculate as to the nature of extraterrestrials, we can draw on behavioral studies to understand humans. There are many prognoses concerning the results of contact, ranging from little or no effect to dramatic, fundamental changes in almost every aspect of human existence. In this discussion, too, an optimistic theme prevailed. To be sure, not every scenario is positive, and even under positive scenarios some people are likely to react adversely. Nevertheless, the threatening scenarios seem less probable than the benign scenarios, and people will have many resources to help them cope with ET.

Near-Term Consequences

Our initial reactions, which will follow close on the heels of the announcement, will depend on the nature of ET, the information available to us, media coverage, personal predispositions, and social pressures. Our analysis focused on three variables and five prototypes. The three variables are the immediacy or closeness of the alien civilization, their perceived capabilities, and their perceived intentions. The prototypes consisted of five historical precedents where large numbers of people believed that humans had discovered—or had been discovered by—extraterrestrial beings.

Widespread panicky reactions were specific to Orson Welles's *War of the Worlds* broadcast and did not follow the reports of bat men on the moon, the canals on Mars, the discoveries of quasars and pulsars, or, for that matter, UFO sightings and abductions. A SETI-like contact scenario, which involves intercepting a microwave transmission from a capable but probably benign or even friendly civilization light-years away, should be less likely to trigger panic.

People have many lines of defense between learning of a startling discovery and disintegrating under stress. The first line of defense is seeking information to understand the extraterrestrials and the implications of contact for the self. The second line of defense, if necessary, is problem solving, that is, taking protective actions. Finally there are psychological defenses to assuage one's emotions. Only if all these fail will panic result. A close look at listeners' responses to the *War of the Worlds* broadcast shows many sensible and constructive behaviors. In the unlikely event that the aliens turn out to have high capabilities and evil intentions, their remote location should buy us time to find constructive ways to respond.

The initial news is likely to include scant detail about the extraterrestrials and their society. Even if it is possible to decipher their message, it will be incomplete, dated, and possibly misleading. Further omissions and distortions may occur if the discovery is leaked or when it is officially announced, as the media cover the event, and as rumor spreads from mouth to mouth. Incompleteness and ambiguity mean that initial reactions will draw heavily on human expectations, hopes, and fears.

People routinely form detailed impressions based on minimal information. One of the strongest contributors is likely to be their expectations. People will see what they expect to see, whether they expect angels, devils, or something in between. Despite the admonition that "you can't tell a book by its cover," ET's physical appearance, if known, is likely to be given great importance. Impressions will be shaped, further, by one's intelligence and personality, other people's opinions, cultural and historical contexts, and the state of international affairs at the time contact is made.

If we receive a message from the stars, many people will want to respond. Human curiosity, the lure of technological advances and material wealth, and the fear of damage that might be done by an "unauthorized" response are among the factors that will encourage a prompt, formal response. One possibility is that so much distance will separate the two civilizations that they will not be able to enter into meaningful discourse. Another possibility is that the two cultures will be able to interact, either because they will not be all that far from each other, or because a communications system will exist that "beats" the speed of light.

Societies use three basic techniques to influence one another and achieve favorable results for themselves. The first is through the use of intimidation and force. Because extraterrestrial technology is likely to be much more advanced than ours—and because force is unlikely to be attractive to them—force should not be a salient option. Another strategy is trade or exchange. This can be useful for both parties, but because of the difficulties of interstellar travel, it may be based on the trade of information rather than material resources or manufactured goods. The final, or "integrative," strategy is based on the development of shared values and mutual understanding. The diplomatic protocols that humans have proposed for responding to ET call for mutual respect and a policy of noninterference. These protocols bear a clear resemblance to the principles that terrestrial societies use to get along with one another and avoid war.

Long-Term Consequences

In the long run, contact with an extraterrestrial civilization could have a pervasive impact on our philosophy, our arts, and above all our science and technology. They could help us solve our problems and usher us into a new Golden Age. On the other hand, the introduction of their ideas into our society could backfire and create a nightmare without end.

What if the first civilization that we hear from refuses to cooperate, doesn't provide useful information, or can't be understood? The discovery of even one other intelligent civilization would prove that the basic processes that led to life and intelligence on Earth were duplicated elsewhere. After the basic riddle is solved—intelligent life exists elsewhere—the search will continue until we find some friends or join the Galactic Club.

We considered several prototypes for predicting the effects of contact of their "advanced" culture on our "mediocre" culture. One such prototype was the arrival of technologically oriented Europeans in Africa, Asia, and the Americas. The European and non-European cultures were tremendously different; the Europeans brought with them ideas, animals, and artifacts that had not been dreamt of by many of the people they encountered. Despite their self-defined edge in philosophy, religion, and science, the Europeans' arrival did not have the tremendous uplifting and civilizing effect that they had envisioned. At the same time, their arrival was not entirely bereft of benefits for the affected cultures. What emerges is a complex picture of domination, dependence, the introduction and diffusion of new ideas—and many differing interpretations of this picture. On the whole, a large population, a substantial resource base, high technology, and a powerful military make it possible to explore and then to gain dominance over cultures with fewer resources (less matter, less energy, less information), at least for a period of time.

There are many objections to using terrestrial exploration and colonization as a model, perhaps the greatest being that humans and aliens are unlikely to come into physical contact with each other. Because we will be limited to swapping information (or, more likely, to receiving ideas) we might seek as prototypes those occasions in history when, without direct contact between the two peoples, one culture's ideas influenced the philosophy, religion, or science of another. Certainly, intellectual history can provide us with some useful examples. At the same time, this may be too limiting, in that the information we get could also affect the nitty-gritty of our daily lives. Information we receive could enable us to replicate their artifacts, perhaps even their organisms, and to experiment with their technology. With instruction, practice, and time we might be able to do much of what they can do. Through nanotechnology, our renditions of their artifacts could be impossible to tell from the originals, except for the fact that the originals would exist in a different corner of the galaxy.

The SETI scenario offers us many choices. If we have significant worries about our security, then we do not answer their call. If we don't like what they have to say, then we shut off our receiver or, better yet, try to find somebody else. We can make choices that protect us if the hypothesis that they are peaceful and benevolent seems incorrect.

Could we make equally good choices dealing with a society that seems ready, willing, and able to help? If their society is tremendously different from ours, then the solutions they have found to their problems may not be all that helpful for solving our own. Additionally, each society is a dynamic, complex whole. Major changes do not affect only one area; they have consequences that extend into many areas, some of which may not be obvious at first. The point is not that we should steadfastly resist change or adopt paranoid views about all proffered advice, but that if new opportunities arise, we must carefully think them through and try to predict their varied effects.

SEARCH WITHOUT END?

A deep, wide gulf separates people who believe that life exists on many worlds and those who believe that we are unique, and only compelling evidence one way or the other will draw the matter to a conclusion. As years go by without a confirmed success, some optimists may become less optimistic, but there is no real way to prove that ET does *not* exist. There will always be hope that different search strategies, better equipment, and greater determination will produce a favorable outcome. Even as hopes for life elsewhere in our solar system dim, our hopes for life in other solar systems shines ever brighter, and it does not seem likely that SETI will soon peter out.

Steven Dick identifies some of the powerful philosophical differences that separate many of those who believe that we live in an inhabited universe from those who believe we are alone.[8] At the heart of the matter is the question of universal biology. According to his analysis, "the whole thrust of physical science since the seventeenth century scientific revolution has been to demonstrate the role of physical law in the universe.... The question at stake in the extraterrestrial-life debate is whether an analogous 'biological law' reigns throughout the universe." Dick adds, "No matter how much we learn about life on Earth and the physical nature of the universe of which we are a part, the question of biological uniqueness is central to the quest for who we are and what our role in nature may be."[9] The extraterrestrial-life debate, he further notes, is one of the last bastions of anthropocentrism.[10]

Several powerful psychological forces help people maintain the views they hold dear.[11] One of these is selective perception, or focusing on information that is consistent with one's needs and expectations; another is biased assimilation, which involves interpreting and remembering information in such a way that it fits with preexisting beliefs. Research on biased assimilation shows that "in the course of examining empirical evidence, people are likely to accept 'confirming' evidence at face value, while subjecting 'disconfirming' evidence to critical evaluation, and as a result draw undue support for their position."[12]

Common sense suggests that if we presented both optimists and pessimists with mixed evidence (some of which was compatible with their views and some of which supported the views of their opponents) each side might become a little less confident of its own position and a little more open to the views of the opposition. Studies of biased assimilation, however, suggest that as a result of accepting the arguments that favored their own position while rejecting the arguments that favored the opposition, both optimists and pessimists would become *more* convinced that they were right. Thus, people who have sharply opposing views are not likely to be swayed by a debate, except insofar as they can latch on to new ideas that strengthen their preexisting attitudes.

People's past investments, including psychological and emotional investments, also discourage them from embracing new views. People whose scientific or religious beliefs would be undermined by the detection of extraterrestrial intelligence are not likely to be pleased by findings that prove them wrong, and people who have invested heavily in SETI are not likely to be pleased by evidence that they are pursuing a will-o'-the-wisp. Indeed, one possible response to disconfirming evidence is to reaffirm one's initial position and then raise the stakes by rededicating oneself to the search for truth. Self-justification is one mechanism that contributes to this escalation of commitment. This consists of actions to convince oneself and the world that one is competent and rational and that past decisions have been correct. Another mechanism, suggested by prospect theory, suggests that, rather than walking away from a failing project and incurring certain losses, people sometimes choose to raise the stakes, thereby risking even greater losses, but getting at least an outside chance of success.[13]

Of course, scientific procedures are supposed to eliminate subjectivity and protect us from the errors that arise from such processes as selective perception and biased assimilation. At some point, science accumulates enough verifiable evidence that reasonable people have to reach a consensus. For example, as telescopes got better and better, it became less and less possible to sustain belief in the canals on Mars.

Yet it took a long time for some people to abandon their belief in the canals of Mars, and Dick's analysis of this has important implications for SETI. Typically, he notes, the search for extraterrestrial intelligence has stretched our scientific methods to their limit.[14] When we use weak telescopes, or can't quite tell whether or not a neighboring star is perturbed by planets, we are left with ambiguous evidence. Of course, it is ambiguous evidence that encourages us to rely on our preconceptions and biases. If SETI is always at the edge of technology, and we are forever adding new strategies and bringing new equipment on-line before it is fully operational, then in the absence of a confirmed detection the stage may be set for a search without end.

Interest in SETI seems to wax, wane, and then wax again. As this book goes to press, even NASA is getting back into the act. For years, NASA provided low levels of funding for a microwave search. On October 12, 1992, the search began, but funding was withdrawn one year thereafter. In October 1996, Leonard David published an article titled "Looking for ET: NASA joins the search."[15] The article, which included a picture of the star of the movie *ET*, described NASA's Origins Program. This incremental program involves four steps: finding solar systems, imaging solar systems, finding small planets, and then, in perhaps one or two decades, mapping Earthlike planets.

NASA director Daniel Goldin stated, "In the not-too-distant

future we will have the technology needed to image any planets that orbit nearby stars. It may be possible to infer [from] their atmospheres or the color of their oceans whether they are life bearing.... This would change everything—no human thought or endeavor would be untouched by the discovery."[16] Once Earthlike planets are discovered we could send spacecraft, notes Alan Dressler,[17] and sometime around the year 2020, adds David, we should "get ready for the discovery of a lifetime."[18]

NASA's renewed interest in SETI should not be surprising. The agency provided much of the driving force for the extraterrestrial-life debate during the last half of the 20th century and for decades played a major role in piecing together the infrastructure for SETI. It was political pressure rather than scientific reasons that caused the agency to terminate its microwave observation project. One possibility, of course, is that if NASA scientists locate Earthlike planets, there will be renewed enthusiasm for NASA-sponsored microwave observation.

BUILDING SUPPORT FOR SETI

Compared with other aspects of space exploration (such as settling our solar system or traveling beyond it) SETI costs are almost invisible. Nonetheless, any major scientific undertaking can benefit from widespread support. How does one enlist public support for, say, a manned mission to Mars or an enhanced SETI program? Recently, Robert Bell and I have reviewed some of the implications of the literature on persuasion for broadening the base of support for a manned Mars mission, and the points we raise should be equally useful in building support for SETI.[19] At the heart of the matter is our sense that many space enthusiasts are highly skilled when they speak before knowledgeable and involved audiences (in other words, people like themselves), but they do not always make appropriate modifications to their persuasive techniques when they address other audiences.

The dozens of studies we reviewed suggest that knowledgeable and involved audiences are relatively sensitive to the quality of the arguments directed at them, but relatively insensitive to the quality of the spokespersons. For such audiences it is the ideas, not the speakers, that count. Knowledgeable and enthusiastic audiences respond favorably to intense, opinionated language and should be encouraged to draw their own conclusions.

Less knowledgeable and less motivated audiences are more sensitive to the credibility (expertise and trustworthiness) of the speaker and more influenced by the number of arguments that are advanced than by their quality. They rely on their impressions of the speaker because they do not have the background to evaluate the arguments, and they follow the simple rule of thumb that whoever comes up with the most reasons must be right. Although responsive to vivid details, less knowledgeable and involved audiences tend to be put off by opinionated language. The less knowledgeable and involved the audience, the greater the need for repetition of the message.

Researchers now believe that both types of audiences should be exposed to both sides of an issue; formerly, the recommendation was to present opposing views only to knowledgeable audiences. In a sense, a speaker who presents opposing points of view "inoculates" listeners so that they won't capitulate as soon as they hear contrasting views later on. By showing that the speaker is open-minded and fair, acknowledging opposing points of view bolsters the speaker's credibility.

We suggest that some audiences are more important than others. It wouldn't make sense to waste time on people who will never support SETI, or to invest much in people who are already enthusiastic supporters. Persuasion should be aimed at people who are in the "corridor of indifference," that is, people who could be active supporters but who have not yet had a chance to make up their minds.

Similarly, it makes sense to cultivate those people whose opinions are likely to count. We mention three target audiences. One of these is media personnel, a group that was discussed in considerable detail in chapter 8. As noted in that chapter, scientists who want to work effectively with the media must take the science writer's needs into account. High credibility, easy accessibility, and a knowledge of how to work within the abilities and interests not only of the reporter but also of his or her audience all count.

A second crucial audience is children. If the current generation fails to conclude the search, then our children or our children's children will move the search forward. Historically, our schools have given precedence to imparting skills necessary for dealing with the immediacies of everyday life. This makes sense, but we also need to make sure that our schools produce at least a few dreamers who are comfortable with the theoretical and who are willing to wait for positive results. SETI supporters are aware of this need and have developed many programs to engage the interest of children (see chapter 1).

A third target group is opinion leaders. These are the people to whom the rest of us turn for advice and guidance. Opinion leaders tend to be cosmopolitan, innovative people who have successfully integrated individuality with social conformity and who are knowledgeable in a particular domain. They tend to learn more, retain more, and integrate new information better than nonleaders. Opinion leaders are typically active, outgoing people who occupy central and accessible positions in organizations and communities. They rely less on television and more on interactive, informative media such as field-specific journals, magazines, and newsletters. In essence, if we convince an opinion leader, the opinion leader convinces others. Thus, our persuasive efforts are leveraged.

CONCLUSION

Future historians of science, looking back, may conclude that late-20th-century SETI activities ushered in a new era of interdisciplinary collaboration. From the outset, SETI physicists and biologists have urged that philosophers, social scientists, educators, and humanists be involved. Such alliances are not easy to form. Ben Finney[20] recalls C. P. Snow's[21] essay on the two intellectual cultures: scientists and humanists. In Snow's view, scientists are empirical, universalistic, quantitative, precise, and concerned with general themes in nature.

Humanists, and many social scientists, are intuitive, relativistic, qualitative, imprecise, and concerned with specific examples and cases. Scientists tend to see humanists as fuzzy, illogical, and mystical; humanists tend to see scientists as narrow, rigid, and unaware of the broader issues. Each culture is convinced of its own correctness and superiority and wonders about the shortsightedness of the other. SETI has gotten scientists and humanists together by offering a fertile ground for each culture to learn about the other and to come to understand each other's views.

Perhaps the two cultures are based on distinctly different psychological types. Psychiatrist Carl Jung identified fundamental differences in how people know the world, process information, and reach decisions.[22] Jung's theory involves several dichotomies, but crucial for us is his distinction between two different ways of reaching decisions: thinking and feeling. *Thinking* rests on the impersonal processes of reason and logic to reach a decision; *feeling* rests on personal, social, or cultural values. Thinking "seeks rational order and plan according to interpersonal logic"; feeling seeks order and plan "according to harmony among subjective values."[23] Such distinctions are measured by the Myers-Briggs Type Indicator, or MBTI.[24] Over the years, hundreds of studies have used the MBTI, and it is possible to compare the proportions of thinkers and feelers in the two cultures.

About two-thirds of physical scientists, biological scientists, and engineers are "thinking types," who prefer impersonal, logical bases for making decisions. The majority of those in social sciences and humanities are "feeling types," whose decisions rest on broad personal and social values. For example, approximately 76 percent of the computer systems analysts, 71 percent of the chemical engineers, 67 percent of the electrical engineers, 66 percent of the life and physical scientists, and 60 percent of biological scientists whose scores contribute to the MBTI test norms are "thinking." Only 46 percent of the psychiatrists, 45 percent of the social scientists, 44 percent of the psychologists, and 34 percent of the counselors who took part in MBTI studies are classified as "thinking types."

Jung's descriptions of thinking and feeling types seem to apply well to the two cultures. Thinking types view themselves as objective, analytical, critical, and given to logic and reason. They view feeling types as subjective, emotional, fuzzy, disorganized, and lacking in analytic or critical ability. Additionally, thinking types note that feeling types are "thin-skinned" and have trouble accepting criticism. Thinking types find it easier to relate to other thinking types than to feeling types and occasionally hurt other people's feelings without meaning to. Feeling types, on the other hand, take strong pride in their values and their ability to relate to a wide range of other people. Feeling types view themselves as thoughtful, considerate, selfless, and altruistic. To the feeling type, the thinking type seem distant and cold. Feeling types are attuned to others' wishes but are easily hurt when they encounter rejection.

As Isabel Briggs Myers and Mary H. McCaulley point out, the different Jungian types need each other. Even as society rests on a division of labor—different people specializing in different activities to create a strong, integrated whole—an assemblage of different psychological types should strengthen and advance SETI. On the one hand, feelers benefit from the thinkers' ability to analyze, to organize, and to find logical flaws in arguments. Thinking types help feeling types remain within a consistent framework and keep their eyes on the goal. They help feeling types reach difficult decisions and stand firm in the face of opposition. On the other hand, thinking types can benefit from the feeling types' ability to deal with human-relations issues. Feeling types should be well equipped to provide expertise bearing on the social and emotional sides of SETI. Feeling types tend to be good at forecasting how other people will respond emotionally and at conciliation. Additionally, the feeling types' ability to arouse enthusiasm and to teach can build acceptance for the thinkers' ideas.

There are hundreds of billions of stars. Somewhere, light-years from here, the transmission from Arecibo, Puerto Rico, continues its journey from Earth. Although it speeds outward at 186,000 miles per second, it is still many thousands of light-years away from its destination of M13. What are the chances that other transmissions are speeding between star systems or even galaxies? Are any transmissions headed our way? Let's stay tuned and find out!

REFERENCES

CHAPTER ONE: THE ENORMOUS CHALLENGE

1. Frank Drake and Dava Sobel, *Is Anyone Out There? The Scientific Search for Extraterrestrial Intelligence* (New York: Delacorte Press, 1992), pp. 180–185.
2. John Billingham, Roger Heyns, David Milne, Stephen Doyle, Michael Klein, John Heilbron, Michael Ashkenazi, Michael Michaud, and Julie Lutz, eds., *Social Implications of Detecting an Extraterrestrial Civilization: A Report of the Workshops on the Cultural Aspects of SETI* (Palo Alto, Calif.: SETI Institute, 1994).
3. Walter Sullivan, *We Are Not Alone: The Continuing Search for Extraterrestrial Intelligence*, rev. ed. (New York: Dutton, 1993), p. 283.
4. Carl Sagan and Iosef Shklovskii, *Intelligent Life in the Universe* (San Francisco: Holden-Day, 1966), pp. 378–379, 422–432.
5. Philip Morrison, John Billingham, and John Wolfe, eds., *The Search for Extraterrestrial Intelligence*, NASA Special Publication SP-419 (Washington, D.C.: National Aeronautics and Space Administration, 1977).
6. Billingham *et al.*, *Social Implications*, chap. 2, pp. 8–9.
7. Jean Heidmann, "SETI Programmes All Over the World (and Further Out)," *Journal of the British Interplanetary Society*, **48**, 447 (1995).
8. David C. Black and Mark A. Stull, "The Science of SETI," in Morrison *et al. The Search*, pp. 93–120.
9. James Grier Miller, *Living Systems* (New York: McGraw-Hill, 1978).
10. Albert A. Harrison, "Thinking Intelligently about Extraterrestrial Intelligence," *Behavioral Science*, **38**, 189 (1993).

11. Michael A. G. Michaud, "On Communicating with Aliens," *Foreign Service Journal*, **33** (June 1974), p. 33.
12. G. McCain and G. Segal, *The Game of Science*, 2nd ed. (Monterey, Calif: Brooks/Cole, 1973).
13. Albert A. Harrison, "Beyond Earthocentrism: Anthropology on the High Frontier," *Space Power* 3, 345 (1988).
14. Frank White, *The SETI Factor* (New York: Walker and Company, 1990), pp. 154–155.
15. Ian Ridpath, *Messages from the Stars: Communication with Extraterrestrial Life* (New York: Harper and Row, 1978), pp. 139–140.
16. Drake and Sobel, *Is Anybody Out There?* pp. 52–54.
17. Amos Tversky, "Elimination by Aspects: A Theory of Choice," *Psychological Review*, 79, 281 (1972).
18. Sybil P. Parker, *Concise Encyclopedia of Science and Technology* (New York: McGraw-Hill, 1992), p. 1,795.
19. Sagan and Shklovskii, *Intelligent Life in the Universe*, p. 130.
20. Morrison *et al.*, *The Search for Extraterrestrial Intelligence*, p. 3.
21. Bruce Campbell, "Looking for Extrasolar Planets," in Ben Bova and Byron Price, eds., *First Contact: The Search for Extraterrestrial Intelligence* (New York: Plume, 1991), p. 97.
22. A Tutukov, "The Planet around PSR 1829–10," in G. Seth Shostak, ed., *Third Decennial US–USSR Conference on SETI* (San Francisco: Astronomical Society of the Pacific, 1993), pp. 185–194.
23. Campbell, "Looking for Extrasolar Planets," p. 97.
24. *Time*, February 5, 1996, p. 53.
25. Tobias Owen, "The Prevalence of Earth-Like Planets," paper presented at the 47th International Astronautical Conference, Beijing, China, October 1996.
26. Campbell, "Looking for Extrasolar Planets," p. 97.
27. David M. Raup, *Extinction: Bad Genes or Bad Luck?* (New York: W. W. Norton, 1991), p. 1.
28. L. Doyle, C. P. McKay, D. Whitmire, J. Matese, R. Reynolds, and W. Davis, "Astrophysical Constraints on Exobiological Habitats," in G. Seth Shostak, ed., *Third Decennial US–USSR Conference on SETI* (San Francisco: Astronomical Society of the Pacific, 1993), pp. 199–218.
29. Owen, "The Prevalence of Earth-Like Planets."
30. Martyn J. Fogg, *Terraforming: Engineering Planetary Environments* (Warrendale, Penn.: Society of Automotive Engineers, 1995).
31. R. A. Mole, "Terraforming Mars with Four War Surplus Bombs," *Journal of the British Interplanetary Society*, **48**, 321 (1995).
32. Steven J. Dick, *The Biological Universe: The Twentieth Century Extraterrestrial Life Debate and the Limits of Science* (New York: Cambridge University Press, 1996), p. 538.
33. Emmanuel Davoust, *The Cosmic Water Hole* (Cambridge, Mass.: MIT Press, 1991); pp. 1–19; C. Ponnaperuma, Y. Honda, and R. Navarro-Gonzalez, "Chemical Studies on the Existence of Extraterrestrial Life," *Journal of the British Interplanetary Society*, **45**, 241 (1992).
34. Davoust, *The Cosmic Water Hole*, p. 73.
35. Stanley L. Miller and Lesley E. Orgel, *The Origins of Life on Earth* (Englewood Cliffs, NJ: Prentice-Hall, 1974).

36. Raup, *Extinction*, p. 41.
37. Paul Davies, *Are We Alone? Philosophical Implications of the Discovery of Extraterrestrial Life* (New York: Harper Collins, 1995).
38. *Ibid.*, p. 28.
39. Stuart Kauffman, *At Home in the Universe: The Search for Laws of Self-Organization and Complexity* (New York: Oxford University Press, 1995), p. 25.
40. Davies, *Are We Alone?* p. 127.
41. Kauffman, *At Home in the Universe*, p. 43.
42. Dean Falk, "Brain Evolution in Dolphins, Humans, and Other Animals," in G. Seth Shostak, ed., *Progress in the Search for Extraterrestrial Life: 1993 Bioastronomy Symposium* (San Francisco: Astronomical Society of the Pacific, 1995), pp. 53–64.
43. Lori Marino, "SETI Begins at Home: Searching for Terrestrial Intelligence," in Shostak, *Progress in the Search for Extraterrestrial Life*, pp. 545–550.
44. Jean Heidmann, *Extraterrestrial Intelligence* (Cambridge, England: Cambridge University Press, 1995), p. 102.
45. Sagan and Shklovskii, *Intelligent Life in the Universe.*
46. Drake and Sobel, *Is Anyone Out There?* p. 211.
47. Joseph F. Baugher, *On Civilized Stars: The Search for Intelligent Life in Outer Space* (Englewood Cliffs, NJ: Prentice-Hall, 1985), p. 123.
48. Raup, *Extinction*, p. 65.
49. Michael Mautner, "Space-Based Genetic Cryoconservation of Endangered Species," *Journal of the British Interplanetary Society*, **49**, 319 (1996).
50. William Hamilton, "The Discovery of Extraterrestrial Intelligence: A Religious Response," in James L. Christian, ed., *Extraterrestrial Intelligence: The First Encounter* (Buffalo, N.Y.: Prometheus Books, 1976), p. 105.
51. Sagan and Shklovskii, *Intelligent Life in the Universe*, pp. 356–361.
52. Drake and Sobel, *Is Anyone Out There?* p. 52.
53. Davoust, *The Cosmic Water Hole*, p. 118.
54. Quoted in Ian Ridpath, *Messages from the Stars*, pp. 9–10.
55. Quoted in Frank White, *The SETI Factor* (New York: Walker and Company, 1990), p. 196.
56. Drake and Sobel, *Is Anyone Out There?* p. xv.
57. Frank J. Tipler, "Extraterrestrial Beings Do Not Exist," in Edward Regis Jr., ed., *Extraterrestrials: Science and Alien Intelligence* (New York: Cambridge University Press, 1985), p. 147.
58. Baird, *The Inner Limits of Outer Space*, pp. 77–78.
59. Sagan and Shklovskii, *Intelligent Life in the Universe*, p. 12.
60. Daniel Katz, "The Functional Approach to the Study of Attitude Change," *Public Opinion Quarterly*, **24**, 163 (1960).
61. Drake and Sobel, *Is Anyone Out There?* pp. 227–229.
62. Carl Sagan, "The Burden of Skepticism," in Kendrick Frazier, ed., *The Hundredth Monkey and Other Paradigms of the Paranormal* (Buffalo, N.Y.: Prometheus Books, 1991), pp. 151–162.
63. William S. Bainbridge, "Attitudes toward Interstellar Communication: An Empirical Study," *Journal of the British Interplanetary Society*, **36**, 298 (1983).
64. Michael Ruse, "Is Rape Wrong on Andromeda?" in Edward Regis Jr., ed., *Extraterrestrials: Science and Alien Intelligence* (New York: Cambridge University Press, 1985), p. 71.

65. Frank White, *The Overview Effect: Space Exploration and Human Evolution* (Boston: Houghton-Mifflin 1987); *The SETI Factor.*
66. Allen Tough, "Positive Consequences of SETI before Detection," paper presented at the 46th International Astronautical Congress, Oslo, Norway, October 1995.
67. Albert A. Harrison and Alan C. Elms, "Psychology and the Search for Extraterrestrial Intelligence," *Behavioral Science*, **35**, 207 (1990).
68. Jacques Vallee, "The Potential of SETI for Major Existential Models," *Journal of the British Interplanetary Society*, **49**, 283 (1996).
69. Andrew Fraknoi, "E.T., Klingons, and the Galactic Library: SETI and Science Education," in Shostak, *Progress in the Search for Extraterrestrial Life*, pp. 535–536.
70. Ragbir Bhathal, "SETI Australia," paper presented at the 47th International Astronautical Congress, Beijing, China, October 1996.
71. Steven Gillett, *World Building: A Writer's Guide to Constructing Star Systems and Life-Supporting Planets* (Cincinnati, Ohio: Writer's Digest Books, 1996).
72. M. Sherif, O. J. Harvey, B. White, W. Hood, and C. Sherif, *Group Conflict and Cooperation: Their Social Psychology* (London: Routledge and Kegan Paul, 1967).
73. Karl S. Guthke, *The Last Frontier: Imagining Other Worlds from the Copernican Revolution to Modern Science Fiction* (Ithaca, N.Y.: Cornell University Press, 1990); Dick, *The Biological Universe.*
74. Ronald D. Brown, "Opening Remarks," in Shostak, *Progress in the Search for Extraterrestrial Life*, pp. 9–14.
75. Ben R. Finney, "SETI and the Two Terrestrial Cultures," *Acta Astronautica*, **46**, 263 (1992).
76. David W. Swift, *SETI Pioneers* (Tucson, Ariz.: University of Arizona Press, 1990).

CHAPTER TWO: LISTENING

1. James L. Christian, ed., *Extraterrestrial Intelligence: The First Encounter* (Buffalo, N.Y.: Prometheus Books, 1976); M. Maruyama and A. Harkins, eds., *Cultures beyond Earth: The Role of Anthropology in Outer Space* (New York: Vintage Books, 1975).
2. Lawrence M. Krauss, *The Physics of* Star Trek (New York: Basic Books, 1995), p. 25.
3. Ronald W. Bracewell, *The Galactic Club: Intelligent Life in Outer Space* (San Francisco: San Francisco Book Company, 1976), p. 110.
4. Krauss, *The Physics of* Star Trek, p. 25.
5. Robert L. Forward, "21st Century Space Propulsion," *The Journal of Practical Applications in Space*, **2**, 1 (1991).
6. Frank J. Tipler, "Traveling to the Other Side of the Universe," *Journal of the British Interplanetary Society*, **49**, 313 (1996).

7. Nick Herbert, *Faster Than Light: Superluminal Loopholes in Physics* (New York: Plume, 1989).
8. Robert L. Forward, "Ad Astra!" *Journal of the British Interplanetary Society*, **49**, 23 (1996).
9. Eric M. Jones and Ben R. Finney, "Fastships and Nomads: Two Routes to the Stars," in Ben R. Finney and Eric M. Jones, eds., *Interstellar Migration and the Human Experience*, (Berkeley, Calif.: University of Calfornia Press, 1984), pp. 88–104.
10. Finney and Jones, eds., *Interstellar Migration and the Human Experience*.
11. Robert Zubrin, "Detection of Extraterrestrial Civilizations via the Spectal Signature of Advanced Interstellar Spacecraft," *Journal of the British Interplanetary Society*, **49**, 297 (1996).
12. Paul Halpern, *Cosmic Wormholes: The Search for Interstellar Shortcuts* (New York: Dutton, 1992).
13. Claudio Maccone, "Interstellar Travel through Magnetic Wormholes," *Journal of the British Interplanetary Society*, **48**, 453 (1995).
14. Frank Drake and Dava Sobel, *Is Anyone Out There? The Scientific Search for Extraterrestrial Intelligence* (New York: Delacorte Press, 1992), pp. 178–190.
15. B. W. Augenstein, "Interstellar Communication by Calling Card," *Journal of the British Interplanetary Society*, **43**, 235 (1990); Allen Tough, "Detection of Miniature Probes: A Need for a New Protocol?" paper presented at the 47th International Astronautical Congress, Beijing, China, October 1996.
16. Mark J. Carlotto and Michael C. Stein, "A Method for Searching for Artificial Objects on Planetary Surfaces," *Journal of the British Interplanetary Society*, **43**, 209 (1990).
17. Robert A. Freitas, "The Search for Extraterrestrial Artifacts (SETA)," *Journal of the British Interplanetary Society*, **36**, 501 (1983).
18. Tough, "Detecting Miniature Probes."
19. Tipler, "Traveling to the Other Side of the Universe."
20. Jack A. Adams, *Human Factors Engineering* (New York: Macmillan, 1989), p. 77.
21. Steven J. Dick, *The Biological Universe: The Twentieth Century Extraterrestrial Life Debate and the Limits of Science* (New York: Cambridge University Press, 1996).
22. Jill Tarter, "Searching for Extraterrestrials," in Edward Regis Jr., ed., *Extraterrestrials: Science and Alien Intelligence* (New York: Cambridge University Press, 1985), pp. 166–199.
23. Jean Heidmann, "SETI Programmes All Over the World (and Further Out)," *Journal of the British Interplanetary Society*, **48**, 447 (1995).
24. John Billingham, Roger Heyns, David Milne, Stephen Doyle, Michael Klein, John Heilbron, Michael Ashkenazi, Michael Michaud, and Julie Lutz, eds., *Social Implications of Detecting an Extraterrestrial Civilization: A Report of the Workshops on the Cultural Aspects of SETI* (Palo Alto, Calif.: SETI Institute, 1994), chap. 1, p. 5.
25. Carl Sagan and Iosef Shklovskii, *Intelligent Life in the Universe* (San Francisco: Holden-Day, 1966).
26. Jerald Greenberg and Robert A. Baron, *Behavior in Organizations*, 4th ed., (Boston: Allyn and Bacon, 1993), pp. 541–547.
27. Emmanuel Davoust, *The Cosmic Water Hole* (Cambridge, Mass.: MIT Press, 1991), p. 147.

28. William Poundstone, *Big Secrets* (New York: William Morrow and Company, 1983), p. 198.
29. Philip Hough and Jenny Randles, *Looking for the Aliens: A Psychological, Scientific, and Imaginative Investigation* (London: Blandford, 1991), pp. 128–129.
30. *Ibid.*, p. 126.
31. Drake and Sobel, *Is Anyone Out There?* p. 38.
32. Water Sullivan, *We Are Not Alone: The Continuing Search for Extraterrestrial Intelligence*, rev. ed. (New York: Dutton, 1993), p. 214.
33. G. Seth Shostak, ed., *Third Decennial US–USSR Conference on SETI* (San Francisco, Astronomical Society of the Pacific, 1993); G. Seth Shostak, ed., *Progress in the Search for Extraterrestrial Life: 1993 Bioastronomy Symposium* (San Francisco: Astronomical Society of the Pacific, 1995).
34. Krauss, *The Physics of* Star Trek p. 130.
35. Poundstone, *Big Secrets*, p. 199.
36. Drake and Sobel, *Is Anyone Out There?* pp. 229–232.
37. Billingham *et al.*, *Social Implications*, chap. 2, p. 4.
38. J. Tarter, P. Backus, G. Heligman, J. Draher, S. Laroque, and the Project Phoenix Team, "Studies of Radio Frequency Interference at Parkes Observatory," paper presented at the 47th International Astronautical Congress, Beijing, China, October 1996.
39. Claudio Maccone, "The SETISAIL Project," in Shostak, *Progress in the Search for Extraterrestrial Life*, pp. 407–418.
40. Zubrin, "Detection of Extraterrestrial Civilizations."
41. Forward, "Ad Astra!"
42. John Baird, *The Inner Limits of Outer Space* (Hanover, N.H.: University Press of New England, 1987), pp. 148–150, 194–257.
43. Douglas Raybeck, "Problems in Extraterrestrial Communication," paper presented at CONTACT IX, Palo Alto, Calif., March 1992.
44. Baird, *The Inner Limits of Outer Space*, pp. 158–175.
45. Ian Ridpath, *Messages from the Stars: Communication and Contact with Extraterrestrial Life* (New York: Harper & Row), pp. 103–104.
46. Douglas Vakoch, "Possible Pictorial Messages for Communication with Extraterrestrial Intelligence," *Journal of the Minnesota Academy of Science*, **44**, 23 (1978); "An Iconic Approach to Communicating Chemical Concepts to Extraterrestrials," *SPIE Proceedings*, 2704 (1996); "Constructing Messages to Extraterrestrials: An Exosemiotic Approach," *Acta Astronautica*, in press.
47. Raybeck, "Problems in Extraterrestrial Communication."
48. Charles C. Adams, *Boontling: An American Lingo* (Philo, Calif.: Mountain House Press, 1990).
49. Ben R. Finney and Jerry Bentley, "SETI Analogies: Learning at a Distance from Ancient Terrestrial Civilizations," paper presented at the 45th congress of the International Astronautical Federation, Jerusalem, Israel, October 1994.
50. James R. Lewis, ed., *The Gods Have Landed: New Religions from Other Worlds* (Albany, N.Y.: State University of New York Press, 1995).
51. Adrian Berry, *The Next 500 Years: Life in the Coming Millennium* (New York: W. H. Freeman and Company, 1995), p. 255.
52. Vakoch, "Iconic Approach."

53. Hans Freudenthal, "Excerpts from Lincos: Design of a Language for Cosmic Intercourse," in Edward Regis Jr., ed., *Extraterrestrials: Science and Alien Intelligence* (New York: Cambridge University Press, 1985), pp. 215–228.
54. C. L. DeVito and R. T. Oehrle, "A Language Based on the Fundamental Facts of Science," *Journal of the British Interplanetary Society*, **43**, 561 (1990).
55. *Ibid.*, p. 561.
56. Raybeck, "Problems in Extraterrestrial Communication."

CHAPTER THREE: FALSE ALARMS

1. *San Francisco Chronicle*, June 27, 1947, p. 24.
2. Albert A. Harrison and James M. Thomas, "The Kennedy Assassination, Unidentified Flying Objects, and Other Conspiracies: Psychological and Organizational Factors in the Perception of 'Cover-Up,'" *Systems Research and Behavioral Science*, 14, 113 (1997).
3. Curtis Peebles, *Watch the Skies! A Chronicle of the Flying Saucer Myth* (Washington, D.C.: Smithsonian Institution Press, 1994).
4. *Ibid.*, pp. 71–72.
5. J. G. Fuller, *The Interrupted Journey* (New York: Putnam, 1966).
6. Robert L. Hall, Mark Rodeghier and Donald A. Johnson, "The Prevalence of Abductions: A Critical Look," *Journal of UFO Studies*, 4, 131 (1992).
7. David M. Jacobs, *Secret Life: Firsthand Accounts of UFO Abductions* (New York: Simon and Schuster, 1992).
8. Jacobs, *Secret Life*; Philip J. Klass, *UFO Abductions: A Dangerous Game* (Buffalo, N.Y.: Prometheus Books, 1989); Leonard S. Newman and Roy F. Baumeister, "Toward an Explanation of the UFO Abduction Phenomenon: Hypnotic Elaboration, Extraterrestrial Sadomasochism, and Spurious Memories," *Psychological Inquiry*, **7**, 99 (1996).
9. Rene Spitz, "Hospitalism: An Enquiry into the Genesis of Psychiatric Conditions in Early Childhood," *Psychoanalytic Study of the Child*, **1**, 53 (New York: International Universities Press, 1945).
10. Newman and Baumeister, "Toward an Explanation of the UFO Abduction Phenomenon."
11. Jacobs, *Secret Life*, p. 9.
12. Richard F. Haines, *Observing UFOs: An Investigation Handbook* (Chicago: Nelson-Hall, 1979), p. 148.
13. Jacobs, *Secret Life*, pp. 259–260.
14. Harrison and Thomas, "The Kennedy Assassination."
15. Newman and Baumeister, "Toward an Explanation of the UFO Abduction Phenomenon".

16. Hilary Evans, "A Postscript to Ring, Rosing and Baker," *Journal of UFO Studies*, **3**, 133 (1992).
17. Robert A. Baker, "The Aliens among Us: Hypnotic Regression Revisited," in Kendrick Frazier, ed., *The Hundredth Monkey and Other Paradigms of the Paranormal* (Buffalo, N.Y.: Prometheus Books, 1991), pp. 44–69.
18. Klass, *UFO Abductions*, pp. 52–56.
19. *Ibid.*, p. 42.
20. Vincente-Juan Ballester Olmos, "Alleged Experiences inside UFOs: An Analysis of Abduction Reports," *Journal of Scientific Exploration*, **8**, 91 (1994).
21. Muzafer Sherif, *The Psychology of Social Norms* (New York: Harper and Row, 1936).
22. *Washington Post*, June 29, 1947, p. 3.
23. Klass, *UFO Abductions*, p. 13.
24. Haines, *Observing UFOs*, pp. 191–195.
25. Baker, "The Aliens among Us."
26. Nicholas P. Spanos, Cheryl A. Burgess, and Melissa Faith Burgess, "Past Life Identities, UFO Abductions and Satanic Ritual Abuse: The Social Construction of Memories," *International Journal of Clinical and Experimental Hypnosis*, **42**, 433 (1994).
27. Steven E. Clark and Elizabeth Loftus, "The Construction of Space Alien Abduction Memories," *Psychological Inquiry*, **7**, 140 (1996).
28. Robert B. Zajonc, *Social Psychology: An Experimental Approach*, (Belmont, Calif.: Brooks/Cole, 1966).
29. Spanos, Burgess, and Burgess, "Past Life Identities."
30. Bette L. Bottoms, Philip R. Shaver, and Gail Goodman, "Profile of Ritual and Religious-Related Abuse Allegations Reported to Clinical Psychologists in the United States," paper presented at the 99th Annual Convention of the American Psychological Association, San Francisco, August 1991.
31. Keith Thompson, *Angels and Aliens: UFOs and the Mythic Imagination* (Reading, Mass: Addison-Wesley, 1991), p. 60.
32. Newman and Baumeister, "Toward an Explanation of the UFO Abduction Phenomena."
33. Nicholas P. Spanos, Patricia A. Cross, Kirby Dickson, and Susan C. DuBriel, "Close Encounters: An Examination of UFO Experiences," *Journal of Abnormal Psychology*, **102**, 624 (1993).
34. Kenneth H. Ring, *The Omega Project: Near Death Experiences, UFO Encounters, and Mind at Large* (New York: William Morrow and Company, 1992); Mark Rodeghier, Jeff Goodpaster, and Sandra Blatterbauer, "Psychosocial Characteristics of Abductees: Results from the CUFOS Abduction Project," *Journal of UFO Studies*, **3**, 59 (1991).
35. James E. Oberg, *UFOs and Outer Space Mysteries* (Norfolk, Va.: Donning, 1982), pp. 103–107.
36. Jacques Vallee, *Revelations: Alien Contact and Human Deception* (New York: Ballantine Books, 1991), p. 4–5.
37. Jamie Arndt and Jeff Greenberg, "Fantastic Accounts Can Take Many Forms: False Memory Construction? Yes. Escape from Life? We Don't Think So," *Psychological Inquiry*, **7**, 127 (1996).
38. Philip Hough and Jenny Randles, *Looking for the Aliens: A Psychological Scien-*

tific, and Imaginative Investigation (London: Blandford, 1991), pp. 40–43.
39. Jacques Vallee, "Anatomy of a Hoax: The Philadelphia Experiment Fifty Years Later," *Journal of Scientific Exploration*, **8**, 47 (1994).
40. Oberg, *UFOs and Outer Space Mysteries*, p. 116.
41. June O. Parnell and R. Leo Sprinkle, "Personality Characteristics of Persons Who Claim UFO Experiences," *Journal of UFO Studies*, **2**, 45 (1990); Rodeghier *et al.*, "Psychosocial Characteristics of Abductees"; Spanos *et al.*, "Close Encounters."
42. Gregory L. Little, "Educational Level and Primary Beliefs about Unidentified Flying Objects Held by Recognized UFOlogists," *Psychological Reports*, **54**, 907 (1984).
43. Troy A. Zimmer, "Belief in UFOs as Alternative Reality, Cultural Rejection, or Disturbed Psyche," *Deviant Behavior*, **6**, 405 (1985).
44. Spanos *et al.*, "Close Encounters."
45. *Ibid.*, p. 631.
46. Baker, "The Aliens among Us."
47. Kenneth H. Ring, *The Omega Project: Near Death Experiences, UFO Encounters, and Mind at Large*.
48. Susan M. Powers, "Dissociation in Alleged Extraterrestrial Abductees," *Dissociation*, **7**, 44 (1991).
49. Ring, *The Omega Project*.
50. Michael A. Persinger, "Geophysical Variables and Behavior: LV: Predicting Details of Visitor Experiences and the Personality of Experients," *Psychological Reports*, **68**, 55 (1989).
51. For example, J. S. Derr and Michael A. Persinger, "Luminous Phenomena and Seismic Energy in the Central U.S.," *Journal of Scientific Exploration*, **4**, 55 (1990); Michael A. Persinger, "Religious and Mystical Experiences as Artifacts of the Temporal Lobe Function: A General Hypothesis," *Perceptual and Motor Skills*, **57**, 1255 (1983); *The Neurophysiology of God Beliefs* (New York: Praeger, 1987); Michael A. Persinger and J. S. Derr, "Geophysical Variables and Behavior: LXII: Temporal Coupling of UFO Reports and Seismic Energy Release within the Rio Grande Rift System: Discriminative Validity of Tectonic Strain Theory," *Perceptual and Motor Skills*, **71**, 567 (1990).
52. Vallee, *Revelations*, pp. 3–5.
53. Jacques Vallee, *Dimensions: A Casebook of Alien Contact*, (New York: Ballantine Books, 1988).
54. Keith Thompson, *Angels and Aliens: UFOs and the Mythic Imagination* (Reading, Mass.: Addison-Wesley, 1991), p. xi.
55. *Ibid.*, p. 135.
56. Quoted in Ralph Blum and Judy Blum, *Beyond Earth* (New York: Bantam Books, 1974), p. 5.
57. Peter A. Sturrock, "Report on a Survey of the Membership of the American Astronomical Society concerning the UFO Problem, Part I," *Journal of Scientific Exploration*, **8**, 1 (1994).
58. David W. Swift, *SETI Pioneers* (Tucson, Ariz.: University of Arizona Press, 1990).
59. Donald E. Tarter, "Treading on the Edge: Practicing Safe Science with SETI," *Skeptical Inquirer*, **17**, 288 (1993).

CHAPTER FOUR: LIVING SYSTEMS

1. Adrian Berry, *The Next 500 Years: Life in the Coming Millennium* (New York: W. W. Freeman and Company, 1996), p. 296.
2. Emmanuel Davoust, *The Cosmic Water Hole* (Cambridge, Mass.: MIT Press, 1991), pp. 4–5.
3. Carl Sagan and Iosef Shklovskii, *Intelligent Life in the Universe* (San Francisco: Holden-Day, 1966), pp. 207–212.
4. Fred Hoyle and Chandra Wickramasinghe, *Our Place in the Cosmos* (London: J. M. Dent, 1993).
5. *Ibid.*, p. 4.
6. Michael Mautner, "Directed Panspermia: Part 2; Technical Advances toward Seeding Other Solar Systems and the Foundation of Panbiotic Ethics," *Journal of the British Interplanetary Society*, **48**, 435 (1995).
7. Christian de Duve, *Vital Dust* (New York: Basic Books, 1995), pp. 6–7.
8. David M. Raup, *Extinction: Bad Genes or Bad Luck?* (New York: E. E. Norton, 1991), p. 3.
9. John S. Kennedy, *The New Anthropomorphism* (New York: Cambridge University Press, 1992).
10. Donald R. Griffin, *Animal Minds* (Chicago: University of Chicago Press, 1992), p. 18.
11. *Ibid.*, p. 24.
12. George Burghardt, "Cognitive Ethology and Critical Anthropomorphism: A Snake with Two Heads and Hog-Nosed Snakes That Play Dead," in Carolyn L. Ristau, ed., *Cognitive Ethology: The Minds of Other Animals* (Hillsdale, N.J.: Lawrence Erlbaum Associates, 1991), pp. 53–90.
13. Albert A. Harrison, "Thinking Intelligently about Extraterrestrial Intelligence," *Behavioral Science*, **38**, 189 (1993).
14. Frank J. Tipler, *The Physics of Immortality* (New York: Doubleday, 1994), p. 1.
15. Barry Beyerstein, "The Brain and Consciousness: Implications for Psi Phenomena," in Kendrick Frazier, ed., *The Hundredth Monkey and Other Paradigms of the Paranormal* (Buffalo, N.Y.: Prometheus Books, 1991), pp. 43–53.
16. John W. Berry, Ype H. Poortinga, Marshall H. Segall, and Pierre R. Dasen, *Cross-Cultural Psychology*, (New York: Cambridge University Press, 1992), p. 4.
17. *Ibid.*, p. 42.
18. *Ibid.*, pp. 256–260.
19. James Grier Miller, *Living Systems* (New York: McGraw-Hill 1978).
20. Davoust, *The Cosmic Water Hole*, p. 72.
21. George A. Seielstad, *At the Heart of the Web: The Inevitable Genesis of Intelligent Life* (Boston: Harcourt Brace Jovanovich, 1989), p. 100.
22. *Ibid.*, p. 15.
23. *Ibid.*, pp. 140–141.

24. Ronald W. Bracewell, *The Galactic Club: Intelligent Life in Outer Space* (San Francisco: San Francisco Book Company, 1976), pp. 80–81.
25. Mary M. Connors, Albert A. Harrison, and Joshua E. Summit, "Crew Systems: Integrating Human and Technical Subsystems for the Exploration of Space," *Behavioral Science*, **39**, 183 (1994).
26. William I. McLaughlin, "Pathways of Evolution for Man and Machine," *Journal of the British Interplanetary Society*, **36**, 215 (1983).
27. Davoust, *The Cosmic Water Hole*, p. 5.
28. Steven J. Dick, *The Biological Universe: The Twentieth Century Extraterrestrial Life Debate and the Limits of Science* (New York: Cambridge University Press, 1996), p. 372.
29. Robert A. Freitas, "The Search for Extraterrestrial Artifacts (SETA)," *Journal of the British Interplanetary Society*, **36**, 501 (1983).
30. Robert Zubrin, "Detection of Extraterrestrial Civilizations via the Spectral Signature of Advanced Interstellar Spacecraft," *Journal of the British Interplanetary Society*, **49**, 297 (1996).
31. Frank Drake and Dava Sobel, *Is Anyone Out There? The Scientific Search for Extraterrestrial Intelligence* (New York: Delacorte Press, 1992), p. 160.
32. Albert A. Harrison, Yvonne A. Clearwater, and Christopher P. McKay, eds., *From Antarctica to Outer Space: Life in Isolation and Confinement* (New York: Springer-Verlag, 1991).
33. Martyn J. Fogg, "Temporal Aspects of the Interaction among the First Galactic Civilizations: The 'Interdict Hypothesis,'" *Icarus*, **69**, 370 (1987); "The Feasibility of Intergalactic Colonization and Its Relevance to SETI," *Journal of the British Interplanetary Society*, **41**, 491 (1988).
34. Peter M. Molton, "An Experimental Approach to Extraterrestrial Life," *Journal of the British Interplanetary Society*, **42**, 423 (1989).
35. Karl S. Guthke, *The Last Frontier: Imagining Other Worlds from the Copernican Revolution to Modern Science Fiction* (Ithaca, N.Y.: Cornell University Press, 1990), p. 22.
36. Dick, *The Biological Universe*, p. 222

CHAPTER FIVE: ORGANISMS

1. Frank Drake and Dava Sobel, *Is Anyone Out There? The Scientific Search for Extraterrestrial Intelligence* (New York: Delacorte Press, 1992), p. xi.
2. John Barrow, *The Artful Universe* (Cambridge, England: Clarendon Press, 1995), pp. 64–65.
3. David G. Blair, "The Interstellar Contact Channel Hypothesis: When Can We Expect to Receive Beacons?" in G. Seth Shostak, ed., *Progress in the Search for*

Extraterrestrial Life: 1993 Bioastronomy Symposium (San Francisco: Astronomical Society of the Pacific, 1995), pp. 268–269.

4. Michael Tooley, "Would ETI's Be Persons?" in James L. Christian, ed., *Extraterrestrial Intelligence: The First Encounter* (Buffalo, N.Y.: Prometheus Books, 1976), pp. 129–146.
5. David Swift, *SETI Pioneers* (Tucson, Ariz.: University of Arizona Press, 1990), p. 273.
6. Ronald W. Bracewell, *The Galactic Club: Intelligent Life in Outer Space* (San Francisco: San Francisco Book Company, 1976), pp. 14–15; Emmanuel Davoust, *The Cosmic Water Hole* (Cambridge, Mass.: MIT Press, 1991), pp. 72–73; Hal Clement, "Alternative Life Designs," in Ben Bova and Byron Price, eds., *First Contact: The Search for Extraterrestrial Intelligence* (New York, Plume, 1991), pp. 41–54.
7. Clement, "Alternative Life Designs," p. 46.
8. Davoust, *The Cosmic Water Hole*, p. 73.
9. George A. Seielstad, *At the Heart of the Web: The Inevitable Genesis of Intelligent Life* (Boston: Harcourt Brace Jovanovich, 1989), p. 54.
10. Barrow, *The Artful Universe*, pp. 65–66.
11. Carl Sagan and Iosef Shklovskii, *Intelligent Life in the Universe* (San Francisco, Holden-Day, 1966), p. 350.
12. Barrow, *The Artful Universe*, p. 64.
13. Martin Gardner, *The New Ambidextrous Universe: Symmetry and Asymmetry from Mirror Reflections to Superstrings*, 3rd ed. (New York: W. H. Freeman and Company, 1990), pp. 53–54.
14. Joseph Royce, "Consciousness in the Cosmos," in James L. Christian, ed., *Extraterrestrial Intelligence: The First Encounter* (Buffalo, N.Y.: Prometheus Books, 1976), pp. 177–196.
15. Seielstad, *At the Heart of the Web*, pp. 60–81.
16. Michael H. Hart, "Interstellar Migration, the Biological Revolution, and the Future of the Galaxy," in Ben R. Finney and Eric M. Jones, eds., *Interstellar Migration and the Human Experience* (Berkeley, Calif.: University of California Press, 1984), pp. 278–292.
17. Drake and Sobel, *Is Anyone Out There?*, p. 160.
18. *Ibid.*, p. 160.
19. Alfred W. Crosby, "Life (with All of Its Problems) in Space," in Finney and Jones, eds., *Interstellar Migration and the Human Experience*, pp. 210–219.
20. *Ibid.*
21. Barrow, *The Artful Universe*, pp. 85–86.
22. *Newsweek*, June 14, 1993, p. 58.
23. Ben R. Finney and Eric M. Jones, "Epilog," in Finney and Jones, eds., *Interstellar Migration and the Human Experience*, pp. 333–339.
24. Seielstad, *At the Heart of the Web*, p. 185.
25. Drake and Sobel, *Is Anyone Out There?* p. 68.
26. Richard F. Haines, *Observing UFOs: An Investigative Handbook* (Chicago: Nelson-Hall, 1979), p. 101.
27. John Alcock, *Animal Behavior*, 4th ed. (Sunderland, Mass.: Sinauer Associates, 1989), p. 135.
28. John W. Berry, Ype H. Poortinga, Marshall H. Segall, and Pierre R. Dasen,

Cross-Cultural Psychology (New York: Cambridge University Press, 1992), p. 160.
29. Barrow, *The Artful Universe*, pp. 174–175.
30. Joseph Royce, "Consciousness in the Cosmos," p. 193.
31. Carolyn L. Ristau, "Aspects of the Cognitive Ethology of an Injury-Feigning Bird: The Piping Plover," in Carolyn L. Ristau, ed., *Cognitive Ethology: The Minds of Other Animals* (Hillsdale, N.J.: Lawrence Erlbaum Associates, 1991), p. 119.
32. Alison Jolly, "Conscious Chimpanzees? A Review of Recent Literature," in Ristau, *Cognitive Ethology*, p. 246.
33. Lori Marino, "SETI Begins at Home: Searching for Terrestrial Intelligence," in Shostak, *Progress in the Search for Extraterrestrial Life*, pp. 73–82.
34. Frank J. Tipler, *The Physics of Immortality* (New York: Doubleday, 1994), pp. 22–23.
35. Donald R. Griffin, *Animal Minds* (Chicago: University of Chicago Press, 1992).
36. Ristau, "Aspects of the Cognitive Ethology."
37. Jolly, "Conscious Chimpanzees?"
38. Lori Marino, Diana Reiss, and Gordon G. Gallup, Jr., "Self-Recognition in the Bottlenose Dolphin: A Methodological Test Case for the Study of Extraterrestrial Intelligence," in G. Seth Shostak, ed., *Third Decennial US–USSR Conference on SETI* (San Francisco: Astronomical Society of the Pacific, 1993), pp. 393–402.
39. Douglas K. Candland, *Feral Children and Clever Animals: Reflections on Human Nature* (New York: Oxford University Press, 1993).
40. Irene M. Pepperberg, "A Communicative Approach to Animal Intelligence: A Study of Conceptual Abilities of an African Gray Parrot," in Ristau, *Cognitive Ethology*, pp. 153–186.
41. Dean Falk, "Brain Evolution in Dolphins, Humans, and Other Animals," in Shostak, *Progress in the Search for Extraterrestrial Life*, pp. 53–64.
42. D. A. Russell and R. Seguin, "Reconstruction of the Small Cretaceous Theropod *Stenonychosaurus inequalis* and a Hypothetical Dinosaurid," *Syllogeus*, **37**, 1 (1982).
43. *Ibid.*, p. 34.
44. C. P. McKay, comments at CONTACT XII, Milpitas, Calif., 1995.
45. Howard Gardner, *Frames of Mind: The Theory of Multiple Intelligences* (New York: Basic Books, 1985), pp. 12–29.
46. *Ibid.*, pp. 73–395.
47. Candland, *Feral Children*, pp. 111–133.
48. Royce, "Consciousness in the Cosmos."
49. Timothy Ferris, *The Mind's Sky* (New York: Bantam, 1992), p. 112.
50. Philip Morrison, "Reflections on the Bigger Picture," in Ben Bova and Brian Price, eds., *First Contact: The Search for Extraterrestrial Intelligence* (New York: Plume, 1991), p. 314.
51. Diana Reiss, "The Dolphin: An Alien Intelligence," in Bova and Price, eds., *First Contact*, p. 34.
52. Jean Piaget and B. Inhelder, *The Psychology of the Child* (New York: Basic Books, 1969).
53. John Baird, *The Inner Limits of Outer Space* (Hanover, N.H.: University Press of New England, 1987), pp. 85–102.

54. Margaret Donaldson, *Human Minds: An Exploration* (New York: Allen Lane/ The Penguin Press, 1992).
55. Frank White, *The SETI Factor* (New York: Walker and Company, 1990), p. 15.
56. E. J. Coffey, "The Improbability of Behavioural Convergence in Aliens: Behavioural Implications of Morphology," *Journal of the British Interplanetary Society*, **38**, 515 (1985); "Evolutionary Objection to Alien Design Models," **39**, 508, (1986); "The Anthropic Fallacy," **45**, 23 (1992); "Close Encounters of the Fourth Kind," **45**, 31 (1992).
57. Coffey, "The Improbability of Behavioural Convergence," p. 518.

CHAPTER SIX: SOCIETIES

1. Kenyon B. DeGreene, "Rigidity and Fragility of Large Sociotechnical Systems: Advanced Information Technology," *Behavioral Science*, **36**, 69 (1991).
2. J. W. Deardorff, "Examination of the Embargo Hypothesis as an Explanation for the Great Silence," *Journal of the British Interplanetary Society*, **40**, 373 (1987).
3. Shirley A. Varughese, "The Planet Xeno," in M. Maruyama and A. Harkins, eds., *Cultures beyond Earth: The Role of Anthropology in Outer Space* (New York: Vintage Books, 1975), pp. 129–166.
4. Frank J. Tipler, *The Physics of Immortality* (New York: Doubleday, 1994), pp. 44–55.
5. M. Maruyama and A. Harkins, eds., *Cultures beyond Earth*; J. L. Christian, ed., *Extraterrestrial Intelligence: The First Encounter* (Buffalo, N.Y.: Prometheus Books, 1976).
6. John W. Berry, Ype H. Poortinga, Marshall H. Segall, and Pierre R. Dasen, *Cross-Cultural Psychology* (New York: Cambridge University Press, 1992) pp. 1, 166–167.
7. Paul Bohannon, *How Culture Works* (New York: Free Press, 1995), p. 3.
8. Nancy M. Tanner, "Interstellar Migration: The Beginning of a Familiar Process in a New Context," in Ben R. Finney and Eric M. Jones, eds., *Interstellar Migration and the Human Experience*, (Berkeley, Calif.: University of California Press, 1985), p. 232.
9. Berry *et al.*, *Cross-Cultural Psychology*, pp. 166–167.
10. John D. Barrow, *The Artful Universe* (Oxford, England: Clarendon Press, 1995), p. 82.
11. Sandra L. Bem and Daryl J. Bem, "Case Study of a Non-Conscious Ideology: Training the Woman to Know Her Place," in Daryl J. Bem, ed., *Beliefs, Attitudes and Human Affairs* (Monterey, Calif.: Brooks/Cole, 1970).
12. Bohannon, *How Culture Works*.
13. John Keegan, *A History of Warfare* (New York: Vintage Press, 1994).
14. Stanley Milgram, *Obedience to Authority* (New York: Harper and Row, 1974).

15. Lori Marino, "SETI Begins at Home: The Search for Terrestrial Intelligence", in G. Seth Shostak, ed., *Progress in the Search for Extraterrestrial Life* (San Francisco: Astronomical Society of the Pacific, 1995), pp. 73–81.
16. John Alcock, *Animal Behavior*, 4th ed. (Sunderland, Mass: Sinauer Associates, 1989), p. 522.
17. Erik Erikson, *Childhood and Society* (New York: Norton, 1950).
18. Alcock, *Animal Behavior*, p. 306.
19. Adrian Berry, *The Next 500 Years: Life in the Coming Millennium* (New York: W. H. Freeman and Company, 1996).
20. Vladimir Lytkin, Ben R. Finney, and Liudmila Alepko, "The Planets Are Occupied by Living Beings: Tsiolkovsky, Russian Cosmism, and ETI," paper presented at the 1994 International Conference on SETI and Society, Chamonix, France, June 1994.
21. Marcia Freeman, *How We Got to the Moon: The Story of the German Space Pioneers* (Washington, D.C.: 21st Century Press, 1993); "A Memoir: Krafft Ehricke's Extraterrestrial Imperative," *21st Century Science and Technology*, **4**, 32 (1994).
22. T. A. Heppenheimer, *Colonies in Space* (Harrisburg, Penn.: Stackpole Books, 1977); Gerard K. O'Neal, *The High Frontier* (New York: Bantam Books, 1977); Marshall T. Savage, *The Millennial Project: Colonizing the Galaxy in Eight Easy Steps* (Boston: Little Brown and Company, 1994).
23. O'Neal, *The High Frontier*.
24. Magorah Maruyama, "Designing a Space Community," *The Futurist*, 239 (October 1976).
25. James Grier Miller, "Application of Living Systems Theory to Life in Space," in Albert A. Harrison, Yvonne A. Clearwater, and Christopher P. McKay, eds., *From Antarctica to Outer Space: Life in Isolation and Confinement* (New York: Springer-Verlag, 1991), pp. 177–198.
26. Mary M. Connors, Albert A. Harrison, and Joshua E. Summit, "Crew Systems: Integrating Human and Technical Subsystems for the Exploration of Space," *Behavioral Science*, **49**, 183 (1994).
27. John Ackerman, *To Catch a Falling Star: A Scientific Theory of UFOs* (San Diego: Univelt, 1989).
28. N. S. Karadashev, "Transmission of Information by Extraterrestrial Civilizations," in Donald Goldsmith, ed., *The Quest for Extraterrestrial Life* (Mill Valley, Calif.: University Science Books, 1980), pp. 136–139.
29. Carl Sagan, "On the Detectivity of Advanced Galactic Civilizations," in Goldsmith, ed., *The Quest for Extraterrestrial Life*, pp. 140–141.
30. Freeman Dyson, "Search for Artificial Stellar Sources of Infrared Radiation," in Goldsmith, ed., *The Quest for Extraterrestrial Life*, pp. 108–109.
31. Joseph F. Baugher, *On Civilized Stars: The Search for Intelligent Life in Outer Space* (Englewood Cliffs, N.J.: Prentice-Hall, 1985), p. 116.
32. J. Jugaku, K. Noguchi, and S. Nishimura, "A Search for Dyson Spheres around Late-Type Stars in the Solar Neighborhood," in G. Seth Shostak, ed., *Progress in the Search for Extraterrestrial Life*, pp. 381–386.
33. Donald E. Tarter, "Alternative Models for Detecting Very Advanced Extra-Terrestrial Civilizations," *Journal of the British Interplanetary Society*, **49**, 291 (1996).

34. David H. Freedman, *Brainmakers: How Scientists Are Moving beyond Computers to Create a Rival to the Human Brain* (New York: Simon and Schuster, 1994).
35. Tipler, *The Physics of Immortality*, p. 244.
36. John Baird, *The Inner Limits of Outer Space* (Hanover, NH: University Press of New England, 1987), pp. 112–117.
37. Mary M. Connors, Albert A. Harrison, and Faren R. Akins, *Living Aloft: Human Requirements for Extended Spaceflight* (Washington, D.C.: National Aeronautics and Space Administration, 1985), pp. 265–266.
38. Bruce Russett, *Grasping the Democratic Peace: Principles for a Post-Cold War World* (Princeton, N.J.: Princeton University Press, 1993).
39. Bruce Bueno de Mesquite and Randolph M. Siverson, "War and the Survival of Political Leaders: A Comparative Study of Regime Types and Political Accountability," *American Political Science Review*, **89** 841 (1995).
40. Francis Fukuyama, *The End of History and the Last Man* (New York: Avon Books, 1992).
41. David Blair, "The Interstellar Contact Channel Hypothesis: When Can We Expect to Receive Beacons?" in Shostak, ed., *Progress in the Search for Extraterrestrial Intelligence*, pp. 257–274.
42. John Mueller, *Retreat from Doomsday: The Obsolescence of Major War* (New York: Basic Books, 1988).
43. Berry, *The Next 500 Years*.

CHAPTER SEVEN: SUPRANATIONAL SYSTEMS

1. Richard L. Thompson, *Alien Identities: Ancient Insights into Modern UFO Phenomena* (San Diego: Govardhan Hill, 1993), pp. 211–222, 327.
2. Ronald W. Bracewell, *The Galactic Club: Intelligent Life in Outer Space* (San Francisco: San Francisco Book Company, 1976), pp. 80–87.
3. James Lee Ray, *Global Politics*, 5th ed. (Boston: Houghton-Mifflin, 1992), p. 539.
4. Bracewell, *The Galactic Club*, p. 64.
5. Fred Hoyle and Chandra Wickramasinghe, *Our Place in the Cosmos* (London: J. M. Dent, 1993).
6. George A. Seielstad, *At the Heart of the Web: The Inevitable Genesis of Intelligent Life* (Boston: Harcourt Brace Jovanovich, 1989), pp. 60–81.
7. Claudio Maccone, "SETI via Wormholes," paper presented at the 47th International Astronautical Congress, Beijing, China, October 1996.
8. Nick Herbert, *Faster Than Light: Superluminal Loopholes in Physics* (New York: Plume, 1989).
9. Donald E. Tarter, "Is Real-Time Communication between Distant Civilizations in Space Possible? A Call for Research," paper presented at the International Astronomical Association Symposium, Turin, Italy, June 1996.

10. Bracewell, *The Galactic Club*, p. 53.
11. Timothy Ferris, *The Mind's Sky* (New York: Bantam Books, 1992), p. 31.
12. Bracewell, *The Galactic Club*, p. 33.
13. Ferris, *The Mind's Sky*, pp. 45–54.
14. Clyde Wilcox, "Governing Galactic Civilization: Hobbes and Locke in Outer Space," *Extrapolation*, **32**, 111 (1991).
15. Bruce Russett, *Grasping the Democratic Peace: Principles for a Post-Cold War World* (Princeton, N.J.: Princeton University Press, 1993).
16. John Keegan, *A History of Warfare* (New York: Vintage Books, 1994), pp. 56–60.
17. *Ibid.*, p. 56.
18. John Mueller, *Retreat from Doomsday: The Obsolescence of Modern War* (New York: Basic Books, 1988).
19. Ian A. Crawford, "Space, World Government, and the 'End of History,'" *Journal of the British Interplanetary Society*, **47**, 415 (1994).
20. *Ibid.*
21. Thomas R. Cusack and Richard A. Stoll, "Collective Security and State Survival in the Interstate System," *International Studies Quarterly*, **38**, 33 (1994).
22. *Ibid.*, p. 56.
23. Vladimir Lytkin, Ben R. Finney, and Liudmila Alepko, "The Planets are Occupied by Living Beings: Tsiolkovsky, Russian Cosmism, and ETI," paper presented at the 1994 International Conference on SETI and Society, Chamonix, France, June 1994.
24. *Ibid.*, p. 11.
25. *Ibid.*, p. 12.
26. Robert A. Freitas, "There Is No Fermi Paradox," *Icarus*, **62**, 518 (1985).
27. Frank J. Tipler, "Extraterrestrial Beings Do Not Exist", in Edward Regis Jr., ed., *Extraterrestrials: Science and Alien Intelligence* (New York: Cambridge University Press, 1985), pp. 133–149.
28. Martyn J. Fogg, "Temporal Aspects of the Interaction among the First Galactic Civilizations: The Interdict Hypothesis," *Icarus*, **69**, 370 (1987).
29. M. D. Nussinov and V. I. Maron, "Evolutionary Approach to SETI Problems," *Journal of the British Interplanetary Society*, **46**, 399 (1993).
30. Neil Steinberg, *Complete and Utter Failure* (New York: Doubleday, 1994), p. 50.
31. J. W. Deardorff, "Examination of the Embargo Hypothesis as an Explanation for the Great Silence," *Journal of the British Interplanetary Society*, **40**, 373 (1987).
32. John A. Ball, "The Zoo Hypothesis," *Icarus*, **19**, 347 (1973).
33. Michael H. Hart, "An Explanation for the Absence of Extraterrestrials on Earth," in Ben Zuckerman and Michael H. Hart, eds., *Extraterrestrials: Where Are They?* 2nd ed. (New York: Cambridge University Press, 1996), pp. 1–8.
34. Poul Anderson, "On Fermi's Question," paper presented at CONTACT IV, Sacramento, Calif: March 1987.
35. Cusack and Stoll, *Collective Security and State Survival.*
36. Carl Sagan and Iosef Shklovskii, *Intelligent Life in the Universe* (San Francisco: Holden-Day, 1966), pp. 453–460.
37. Robert A. Freitas, "Extraterrestrial Intelligence in the Solar System: Resolving the Fermi Paradox," *Journal of the British Interplanetary Society*, **36**, 496 (1983); "If They Are Here, Where Are They? Observational and Search Requirements," *Icarus*, **55**, 337 (1983); "There Is No Fermi Paradox."

38. Timothy Good, *Above Top Secret*, (New York: William Morrow and Company, 1988).

CHAPTER EIGHT: FIRST IMPRESSIONS

1. James Axtell, *Beyond 1492: Encounters in Colonial North America* (New York: Oxford University Press, 1992), p. 30.
2. *Ibid.*, p. 31.
3. Fernando Cervantes, *The Devil in the New World: The Impact of Diabolism in New Spain* (New Haven, Conn.: Yale University Press, 1994), p. 8.
4. *Ibid.*
5. John S. Kennedy, *The New Anthropomorphism* (New York: Cambridge University Press, 1992), p. 5.
6. Donald E. Tarter, "Interpreting and Reporting on a SETI Discovery," *Space Policy*, 137 (May 1992).
7. Mary M. Connors, "The Role of the Social Scientist in the Search for Extraterrestrial Intelligence," NASA–Ames Research Center, 1976; "The Consequences of Detecting Extraterrestrial Intelligence for Telecommunication Policy," NASA–Ames Research Center, 1977.
8. Donald E. Tarter, "Reply Policy and Signal Type: Assumptions Drawn from Minimal Source Information," paper presented at the 46th International Astronautical Congress, Oslo, Norway, October 1995.
9. Tarter, "Interpreting and Reporting."
10. Donelson R. Forsythe, *Social Psychology* (Belmont, Calif.: Wadsworth, 1987), p. 404.
11. Ian T. Mitroff and Warren Bennis, *The Unreality Industry* (New York: Birch Lane Press, 1989).
12. David W. Swift, "Responses to Contact: Variables to Consider," in G. Seth Shostak, ed., *Progress in the Search for Extraterrestrial Life: 1993 Bioastronomy Symposium* (San Francisco: Astronomical Society of the Pacific, 1995), pp. 567–572.
13. Frank White, *The SETI Factor* (New York: Walker and Company, 1990), pp. 205–214.
14. Tarter, "Reply Policy and Signal Type."
15. Albert A. Harrison and Robert A. Bell, "Building Support for the Manned Exploration of Mars: Lessons from Theory and Research on Persuasion and Attitude Change," paper presented at Case for Mars VI, Boulder, Colo., July 1996.
16. Albert A. Harrison and James M. Thomas, "The Kennedy Assassination, Unidentified Flying Objects, and Other Conspiracies: Psychological and Organizational Factors in the Perception of 'Cover-Up,'" *Systems Research and Behavioral Science*, 14, 113 (1997).

17. Harrison and Bell, "Building Support for the Manned Exploration of Mars."
18. Irving L. Janis, "Effects of Fear Arousal on Attitude Change: Recent Developments in Theory and Experimental Research," in L. Berkowitz, ed., *Advances in Experimental Social Psychology*, vol. 3 (New York: Academic Press, 1967), pp. 166–224.
19. Dorothy Nelkin, *Selling Science: How the Press Covers Science and Technology*, rev. ed., (New York: W. H. Freeman and Company, 1995).
20. Donald E. Tarter, "SETI and the Media: Views from the Inside Out," paper presented at the 40th Congress of the International Astronautical Federation, Málaga, Spain, 1989.
21. Kendrick Frazier, "First Contact: The News Event and the Human Response," in James L. Christian, ed., *Extraterrestrial Intelligence: The First Encounter* (Buffalo, N.Y.: Prometheus Books, 1975), p. 78.
22. Tarter, "SETI and the Media," 1989.
23. Mitroff and Bennis, *The Unreality Industry*.
24. Terrence O'Flaherty, quoted in Jerry Kroth, *Omens and Oracles: Collective Psychology in the Nuclear Age* (New York: Praeger, 1992), p. 128.
25. Jerry Kroth, *Omens and Oracles: Collective Psychology in the Nuclear Age* (New York: Praeger, 1992), p. 141.
26. Vivian Sobchack, *Screening Space: The American Science Fiction Film* (New York: Ungar, 1987).
27. David O. Sears and Rick Kosterman, "Mass Media and Political Persuasion," in Sharon Shavitt and Timothy A. Brock, eds., *Persuasion: Psychological Insights and Perspectives* (Boston: Allyn and Bacon, 1994), pp. 251–278.
28. Connors, "The Role of the Social Scientist."
29. Joseph Bulgatz, *Ponzi Schemes, Invaders from Mars and More Extaordinary Popular Delusions and the Madness of Crowds* (New York: Harmony Books, 1992), p. 42.
30. M. D. Coovert and G. D. Reeder, "Negativity Effects in Impression Formation: The Role of Unit Formation and Schematic Representations," *Journal of Experimental Social Psychology*, **26**, 49 (1990).
31. Erving Goffman, *Stigma: Notes on the Management of Identity* (Englewood Cliffs, N.J.: Prentice-Hall, 1963).
32. George Burghardt, "Cognitive Ethology and Critical Anthropomorphism: A Snake with Two Heads and Hog-Nosed Snakes That Play Dead," in Carolyn L. Ristau, ed., *Cognitive Ethology: The Minds of Other Animals* (Hillsdale, N.J.: Lawrence Erlbaum Associates, 1991), pp. 53–90.
33. J. B. Birdsell, "Biological Dimensions of Small, Founding Human Populations," in Ben R. Finney and Eric M. Jones, eds., *Interstellar Migration and the Human Experience* (Berkeley, Calif.: University of California Press. 1984), pp. 110–119.
34. Michael J. Lerner, *Belief in a Just World: A Fundamental Delusion* (New York: Plenum, 1980).
35. H. Tajfel and M. Billig, "Familiarity and Categorization in Intergroup Behavior," *Journal of Experimental Social Psychology*, **10**, 159 (1974).
36. J. Harding, H. Proshansky, B. Kutner, and I. Chein, "Prejudice and Ethnic Relations," in G. Lindzey and E. Arsonson, eds., *Handbook of Social Psychology*, 2nd ed., vol. 5 (Reading, Mass.: Addison-Wesley, 1969), pp. 1–76.
37. John Billingham, Roger Heyns, David Milne, Stephen Doyle, Michael Klein, John Heilbron, Michael Ashkenazi, Michael Michaud, and Julie Lutz, eds.,

Social Implications of Detecting an Extraterrestrial Civilization: A Report of the Workshops on the Cultural Aspects of SETI (Palo Alto, Calif.: SETI Institute, 1994).
38. Forsythe, *Social Psychology*, p. 141.
39. Morton Deutsch and Harold B. Gerard, "A Study of Normative and Informational Social Influences upon Individual Judgment," *Journal of Abnormal and Social Psychology*, **51**, 629 (1955).
40. Stanley Schachter, *The Psychology of Affiliation* (Stanford, Calif.: Stanford University Press, 1959).
41. Richard T. Petty, John T. Cacioppo, Alan J. Stratham, and Joseph R. Priester, "To Think or Not to Think: Exploring Two Routes to Persuasion," in Shavitt and Brock, eds., *Persuasion*, pp. 113–147.

CHAPTER NINE: IMPACT

1. Hadley Cantril, *Invasion from Mars* (Princeton, N.J.: Princeton University Press, 1952), pp. 49–53.
2. Mary M. Connors, "The Role of the Social Scientist in the Search for Extraterrestrial Intelligence," NASA–Ames Research Center, 1976.
3. John Billingham, Roger Heyns, David Milne, Stephen Doyle, Michael Klein, John Heilbron, Michael Ashkenazi, Michael Michaud, and Julie Lutz, eds., *Social Implications of Detecting an Extraterrestrial Civilization: A Report of the Workshops on the Cultural Aspects of SETI* (Palo Alto, Calif.: SETI Institute, 1994).
4. Joseph Bulgatz, *Ponzi Schemes, Invaders from Mars and More Extraordinary Popular Delusions and the Madness of Crowds* (New York: Harmony Books, 1992), pp. 142–153.
5. *Ibid.*, p. 145.
6. *Ibid.*, p. 147.
7. William Sheehan, *Planets and Perception: Telescopic Views and Interpretations, 1609–1909* (Tucson, Ariz.: University of Arizona Press, 1988).
8. Karl S. Guthke, *The Last Frontier: Imagining Other Worlds from the Copernican Revolution to Modern Science Fiction* (Ithaca, N.Y.: Cornell University Press, 1990).
9. Sheehan, *Planets and Perception*, p. 169.
10. William E. Burrows, *Exploring Space: Voyages in the Solar System and Beyond* (New York: Random House, 1990), pp. 176–181.
11. Sheehan, *Planets and Perception*, p. 187.
12. Cantril, *Invasion from Mars*.
13. *Ibid.*
14. Burrows, *Exploring Space*, p. 180.
15. Jerry Kroth, *Omens and Oracles: Collective Psychology in the Nuclear Age* (New York: Praeger, 1992), p. 45.

16. Bulgatz, *Ponzi Schemes*, pp. 137–139.
17. *Time*, May 21, 1965, p. 72.
18. *New York Times*, April 13, 1965, p. 29.
19. Sybil P. Parker, *Concise Encyclopedia of Science and Technology* (New York: McGraw-Hill, 1992), p. 1,554.
20. Thomas McDonough, *The Search for Extraterrestrial Intelligence* (New York: Wiley, 1987), p. 125.
21. Walter Sullivan, *We Are Not Alone: The Continuing Search for Extraterrestrial Intelligence*, rev. ed. (New York: Dutton, 1993), p. 210.
22. McDonough, *The Search*, p. 128.
23. Parker, *Concise Encyclopedia*, p. 1,530.
24. Carolyn A. Aldwin, *Stress, Coping, and Development* (New York: The Guilford Press, 1994); Richard S. Lazarus, *Emotion and Adaptation* (New York: Oxford University Press, 1991); Richard S. Lazarus and Calvin Folkman, *Stress, Appraisal, and Coping* (New York: Singer, 1984); Richard S. Lazarus and Bernice N. Lazarus, *Passion and Reason: Making Sense of Our Emotions* (New York: Oxford University Press, 1994).
25. Lazarus and Lazarus, *Passion and Reason*, p. 3.
26. *Ibid.*
27. Daniel E. Berlyne, *Conflict, Arousal and Curiosity* (New York: McGraw-Hill, 1960).
28. Allen Tough, "What Role Will Extraterrestrials Play in Humanity's Future?" *Journal of the British Interplanetary Society*, **39**, 491 (1986).
29. Aldwin, *Stress, Coping, and Development*, pp. 170–190.
30. Erwin R. Parson, "Post-Traumatic Stress Disorder (PTSD): Its Biobehavioral Aspects and Management," in Benjamin B. Wolman and George Stricker, eds., *Anxiety and Related Disorders: A Handbook* (New York: John Wiley and Sons, 1994), pp. 226–288.
31. Aldwin, *Stress, Coping, and Development*, p. 172.
32. Cantril, *Invasion from Mars*.
33. A. Baum, R. Fleming, and J. Singer, "Coping with Victimization by Technological Disaster," *Journal of Social Issues*, **39**, 117 (1983).
34. Marvin Zuckerman, *Behavioral Expressions and Biosocial Bases of Sensation Seeking* (New York: Cambridge University Press, 1994).
35. Arie Kruglanski, *Lay Epistemic Theory in Social–Cognitive Psychology* (New York: Plenum, 1989).
36. Milton Rokeach, *The Open and Closed Mind* (New York: Basic Books, 1960).
37. L. Festinger, H. Riecken, and S. Schachter, *When Prophecy Fails* (Minneapolis: University of Minnesota Press, 1956).
38. Donn Byrne, "Repression–Sensitization as a Dimension of Personality," in B. A. Maher, ed., *Progress in Experimental Personality Research*, vol. 1 (New York: Academic Press, 1964), pp. 169–220.
39. Connors, "The Role of the Social Scientist."
40. Aldwin, *Stress, Coping, and Human Development*, pp. 109–122.
41. Jerald Greenberg and Robert A. Baron, *Behavior in Organizations*, 4th ed. (New York: Allyn and Bacon, 1993), pp. 210–211.
42. Julian B. Rotter, "Some Problems and Misconceptions Related to the Construct

of Internal versus External Control of Reinforcement," *Journal of Consulting and Clinical Psychology*, **43**, 56 (1975); *The Development and Applications of Social Learning Theory: Collected Papers* (New York: Praeger, 1992).
43. Lawrence A. Pervin, *Personality: Theory and Research* (New York: Wiley, 1989), pp. 193–195.
44. John Billingham *et al.*, *Social Implications*, chap. 3, p. 4.
45. David W. Swift, "Responses to Contact: Variables to Consider," in G. Seth Shostak, ed., *Progress in the Search for Extraterrestrial Life: 1993 Biostronomy Symposium* (San Francisco: Astronomical Society of the Pacific, 1995), p. 568.

CHAPTER TEN: BUILDING RELATIONSHIPS

1. Shirley A. Varughese, "The Planet Zeno," in M. Maruyama and A. Harkins, eds., *Cultures beyond Earth: The Role of Anthropology in Outer Space* (New York: Vintage Books, 1975), pp. 164–165.
2. Michael Michaud, "Interstellar Negotiation," *Foreign Service Journal* (December, 1972), pp. 10–14, 29–30; "Negotiating with Other Worlds," *The Futurist*, **7**, 71 (1973).
3. Albert Bandura, *Aggression: A Social Learning Analysis* (New York: Holt, Rinehart and Winston, 1973).
4. Leonard Berkowitz, "Frustration–Aggression Hypothesis: Examination and Reformulation," *Psychological Bulletin*, **106**, 59 (1989).
5. Leonard Berkowitz and Anthony LePage, "Weapons as Aggression-Inducing Stimuli," *Journal of Personality and Social Psychology*, **7**, 202 (1967).
6. Curtis Peebles, *Watch the Skies! A Chronicle of the Flying Saucer Myth* (Washington, D.C.: Smithsonian Institution Press, 1994), pp. 87–102.
7. Ernst Fasan, "Discovery of ETI: Terrestrial and Extraterrestrial Legal Implications," *Acta Astronautica*, **21**, 131 (1990).
8. Stephen E. Doyle, "Post-Detection Global Institutional Arrangements," paper presented at the 47th International Astronautical Congress, Beijing, China, October 1996.
9. *Acta Astronautica*, **21**, 153 (February, 1990).
10. Allan E. Goodman, "Diplomatic and Political Problems Affecting the Formulation and Implementation of an International Protocol for Activities Following the Detection of a Signal for Extraterrestrial Intelligence," *Acta Astronautica*, **21**, 104 (1990).
11. Donald E. Tarter, "Reply Policy and Signal Type: Assumptions Drawn from Minimal Source Information," paper presented at the 46th International Astronautical Congress, Oslo, Norway, October 1995.
12. Allen Tough, "Detecting Miniature Probes: A Need for a New Protocol?" paper

presented at the 47th International Astronautical Conference, Beijing, China, October 1996.
13. Kenneth E. Boulding, *Three Faces of Power* (Newbury Park, Calif.: Sage Publications, 1990), p. 67.
14. Irving L. Janis, *Victims of Groupthink* (Boston: Houghton Mifflin, 1972).
15. John Keegan, *A History of Warfare* (New York: Vintage Books, 1994).
16. Irving L. Janis and Leon Mann, *Decision Making: An Analysis of Conflict, Choice, and Commitment* (New York: Free Press, 1977).
17. Boulding, *Three Faces of Power*.
18. Daniel Headrick, *The Tools of Empire: Technology and European Imperialism in the Nineteenth Century* (New York: Oxford University Press, 1981).
19. Don Marshall, paper presented at CONTACT XII, Milpitas, Calif., 1995.
20. Keegan, *A History of Warfare*, p. 392.
21. Frank White, *The SETI Factor* (New York: Walker and Company, 1990), pp. 154–155.
22. Robert Axelrod, *The Evolution of Cooperation* (New York:Basic Books, 1984).
23. *Ibid.*, p. 190.
24. Boulding, *Three Faces of Power*, p. 3.
25. Michael H. Hart, "Interstellar Migration, the Biological Revolution, and the Future of the Galaxy," in Ben R. Finney and Eric M. Jones, eds., *Interstellar Migration and the Human Experience* (Berkeley, Calif.: University of California Press, 1984), pp. 278–292.
26. Michaud, "Interstellar Negotiation" and "Negotiating with Other Worlds."
27. Barbara D. Moskowitz, "The Moral Obligations of Anthropology," in M. Maruyama and A. Harkins, eds., *Cultures beyond Earth*, pp. 75–76.
28. Ben R. Finney, "The Impact of Contact," *Acta Astronautica*, **21**, 117 (1990).
29. Fasan, "Discovery of ETI."
30. J. Stacey Adams, "The Structure and Dynamics of Behavior in Organizational Boundary Roles," in M. L. Dunnette, ed., *Handbook of Industrial and Organizational Psychology* (Chicago: Rand McNally 1976), pp. 1,125–1,174.
31. Timothy Good, *Above Top Secret* (New York: William Morrow and Company, 1988).
32. Peebles, *Watch the Skies!* p. 112.
33. Allen Tough, "A Critical Examination of Factors that Might Encourage Secrecy," *Acta Astronautica*, **21**, 97 (1990).
34. Carl G. Jung, *Flying Saucers: A Modern Myth of Things Seen in the Skies* (Princeton, N.J.: Princeton University Press, 1959), p. 138.
35. Thomas E. Bullard, "UFO Abduction Reports: The Supernatural Kidnap Narrative Returns in Technological Guise," *Journal of American Folklore*, **102**, 107 (1989).
36. Howard Blum, *Out There* (New York, Pocket Books, 1991), pp. 301–311.
37. Albert A. Harrison and James M. Thomas, "The Kennedy Assassination, Unidentified Flying Objects, and Other Conspiracies: Psychological and Organizational Factors in the Perception of 'Cover-Up,'" *Systems Research and Behavioral Science*, **14**, 113 (1997).
38. Dan Moldea, *The Killing of Robert F. Kennedy: An Investigation of Motive, Means, and Opportunity* (New York, W. W. Norton, 1994).
39. Peebles, *Watch the Skies!* pp. 73–82.

CHAPTER ELEVEN: THE ROCKY ROAD TO UTOPIA

1. Diana Tumminia and R. George Kirkpatrick, "Unarius: Emergent Aspects of an American Flying Saucer Group," in James R. Lewis, ed., *The Gods Have Landed: New Religions from Other Worlds* (Albany, N.Y.: State University of New York Press, 1995), pp. 85–104.
2. Allen Tough, "What Role Will Extraterrestrials Play in Humanity's Future?" *Journal of the British Interplanetary Society*, **39**, 491 (1986).
3. Michael Adas, *Machines as the Measure of Men* (Ithaca, N.Y.: Cornell University Press, 1989).
4. James Axtell, *Beyond 1492: Encounters in Colonial North America*, (New York: Oxford University Press, 1992).
5. Fernando Cervantes, *The Devil in the New World: The Impact of Diabolism in New Spain* (New Haven, Conn.: Yale University Press, 1994).
6. Daniel Headrick, *The Tools of Empire: Technology and European Imperialism in the Nineteenth Century* (New York: Oxford University Press, 1981), p. 208.
7. Steven J. Dick, "Consequences of Success in SETI: Lessons from the History of Science," in G. Seth Shostak, ed., *Progress in the Search for Extraterrestrial Life: 1993 Bioastronomy Symposium* (San Francisco: Astronomical Society of the Pacific, 1995), pp. 521–524.
8. *Ibid.*
9. Frank White, *The SETI Factor* (New York: Walker and Company, 1990), p. 138.
10. Ben R. Finney, "The Impact of Contact," *Acta Astronautica*, **21**, 177 (1990).
11. Arnold Pacey, *The Culture of Technology* (Boston: MIT Press, 1983), p. 147.
12. Robert Axelrod, *The Evolution of Cooperation* (New York: Basic Books, 1984).
13. John D. Barrow, *The Artful Universe* (Oxford, England: Clarendon Press, 1995), p. 39.
14. Adas, *Machines as the Measure of Men*, pp. 357–365.
15. Headrick, *The Tools of Empire*, p. 27.
16. Allen Tough, "A Critical Examination of Factors That Might Encourage Secrecy," *Acta Astronautica*, **21**, 101 (1990).
17. Pacey, *The Culture of Technology*, p. 57.
18. *Ibid.*, p. 10.
19. Finney, "The Impact of Contact."
20. Axtell, *Beyond 1492*, pp. 140–142.
21. Charles Perrow, *Normal Accidents: Living with High-Risk Technologies* (New York: Basic Books, 1984).
22. Axtell, *Beyond 1492*, p. 145.
23. Neil Postman, *Technopoly: The Surrender of Culture to Technology* (New York: Vintage Books, 1993), pp. 28–29.
24. *Ibid.*, p. 18.
25. Rustum Roy, "The Coming Clash of Titans," in Clifford N. Matthews and Roy A. Varghese, eds, *Cosmic Beginnings and Human Ends* (Chicago: Open Court Books, 1995), p. 130.
26. George Bugliarello, "Science at the Crossroads," in Matthews and Varghese, eds., *Cosmic Beginnings and Human Ends*, pp. 109–128.
27. Alvin Toffler, *Future Shock* (New York, Bantam Books, 1970).

28. Roberto Pinotti, "Contact: Releasing the News," *Acta Astronautica*, **21**, 109 (1990).
29. *Ibid.*, p. 110.
30. Michael Fossel, "Resetting the Age Clock," *Stanford Magazine*, 68 (July/August 1996).
31. Frank J. Tipler, *The Physics of Immortality* (New York: Doubleday, 1994), pp. 244–245.
32. Headrick, *The Tools of Empire*, p. 193.
33. Carl G. Jung, *Flying Saucers: A Modern Myth of Things Seen in the Skies* (Princeton, N.J.: Princeton University Press, 1959), p. 134.
34. Mary M. Connors, "The Role of the Social Scientist in the Search for Extraterrestrial Intelligence," NASA–Ames Research Center, 1976.
35. Stuart Kauffman, *At Home in the Universe: The Search for Laws of Self-Organization and Complexity* (New York: Oxford University Press, 1995), pp. 279–280.
36. Pinotti, "Contact," pp. 110–111.
37. Donald E. Tarter, "Reply Policy and Signal Type: Assumptions Drawn from Minimal Source Information," paper presented at the 46th International Astronautical Congress, Oslo, Norway, 1995.
38. Neil Postman, *Technopology: The Surrender of Culture to Technology*.
39. *Ibid.*, p. 53.
40. Karl S. Guthke, *The Last Frontier: Imagining Other Worlds from the Copernican Revolution to Modern Science Fiction* (Ithaca, N.Y.: Cornell University Press, 1990).
41. Ted Peters, "Exotheology: Speculations on Extraterrestrial Life," in James R. Lewis, ed., *The Gods Have Landed: New Religions from Other Worlds* (Albany, N.Y.: State University Press of New York, 1995), pp. 187–206.
42. Michael Ashkenazi, "Not the Sons of Adam: Religious Responses to ETI," paper presented at the 42nd Congress of the International Astronautical Federation, Montreal, Canada, October 1991.
43. Howard Blum, *Out There* (New York: Pocket Books, 1991), pp. 196–213.
44. Philip Morrison, John Billingham, and John Wolfe, eds., *The Search for Extraterrestrial Intelligence*, NASA Special Publication SP-419 (Washington, D.C.: National Aeronautics and Space Administration, 1977), p. vii.
45. Jacques Vallee, "The Potential of SETI for Major Existential Problems," *Journal of the British Interplanetary Society*, **49**, 283 (1996).
46. Dorothy Nelkin, *How the Press Sells Science and Technology*, 2nd ed. (New York: W. H. Freeman and Company, 1995), pp, 32–34.
47. Pacey, *The Culture of Technology*.

CHAPTER TWELVE: BETTING WITH THE OPTIMISTS

1. Paul Tillich, *My Search for Universals* (New York: Simon and Schuster, 1967), pp. 29–30.

2. Herbert S. Jones, *Life on Other Worlds* (New York: Macmillan, 1940), p. 290.
3. Paul Davies, *Are We Alone? Philosophical Implications of the Discovery of Extraterrestrial Life* (New York: Basic Books, 1995), pp. 61–87.
4. John Baird, *The Inner Limits of Outer Space* (Hanover, N.H.: University Press of New England, 1987), p. 133.
5. James Grier Miller, *Living Systems* (New York: McGraw-Hill, 1978).
6. Paul Bohannon, *How Culture Works* (New York: Free Press, 1995), p. 3.
7. Adrian Berry, *The Next 500 Years: Life in the Coming Millennium* (New York: W. H. Freeman and Company, 1996), pp. 9–107.
8. Steven J. Dick, *The Biological Universe* (New York: Cambridge University Press, 1996).
9. *Ibid.*, p. 1.
10. *Ibid.*, p. 51.
11. Albert A. Harrison and James M. Thomas, "The Kennedy Assassination, Unidentified Flying Objects, and Other Conspiracies: Psychological and Organizational Factors in the Perception of 'Cover-Up,'" *Systems Research and Behavioral Science*, **14** 113 (1997).
12. Charles G. Lord, L. Ross, and M. R. Lepper, "Biased Assimilation and Attitude Polarization: The Effects of Prior Theories on Subsequently Considered Evidence," *Journal of Personality and Social Psychology*, **37**, 2098 (1979).
13. G. Whyte, "Escalating Commitment in Individual and Group Decision-Making: A Prospect Theory Approach," *Organizational Behavior and Human Decision Processes*, **54**, 430 (1993).
14. Dick, *The Biological Universe*, p. 543.
15. Leonard David, "Looking for ET: NASA Joins the Search," *Final Frontier*, 24 (October, 1996).
16. *Ibid.*, p. 24.
17. *Ibid.*, p. 29.
18. *Ibid.*, p. 28.
19. Albert A. Harrison and Robert A. Bell, "Building Support for the Manned Exploration of Mars: Lessons from Theory and Research on Persuasion and Attitude Change," paper presented at Case for Mars VI, Boulder, Colo., July 1996.
20. Ben R. Finney, "SETI and the Two Terrestrial Cultures," *Acta Astronautica*, **26**, 263 (1992).
21. C. P. Snow, *The Two Cultures and the Scientific Revolution* (New York: Cambridge University Press, 1959).
22. Carl G. Jung, *Psychological Types* (Princeton, N.J.: Princeton University Press, 1971).
23. Isabel Briggs Myers and Mary H. McCauley, *A Guide to the Development and Use of the Myers-Briggs Type Indicator* (Palo Alto, Calif: Consulting Psychologists Press, 1989), p. 13.
24. *Ibid.*

INDEX